# Collins

# **A2** Revision**Notes**
# Physics

• Ken Price •

Series editor: Jayne de Courcy

William Collins' dream of knowledge for all began with the publication of his first book in 1819. A self-educated mill worker, he not only enriched millions of lives, but also founded a flourishing publishing house. Today, staying true to this spirit, Collins books are packed with inspiration, innovation and practical expertise. They place you at the centre of a world of possibility and give you exactly what you need to explore it.

Collins. Do more.

Published by Collins
An imprint of HarperCollins*Publishers*
77–85 Fulham Palace Road
Hammersmith
London
W6 8JB

Browse the complete Collins catalogue at
**www.collinseducation.com**

© HarperCollins*Publishers* Limited 2006

10 9 8 7 6 5 4 3 2 1

ISBN-13 978 0 00 720688 9
ISBN-10 0 00 720688 7

**British Library Cataloguing in Publication Data**
A Catalogue record for this publication is available from the British Library

Edited by Margaret Shepherd and Mitch Fitton
Production by Katie Butler
Series design by Sally Boothroyd
Illustrated by Kathy Baxendale
Index compiled by Indexing Specialists (UK) Ltd
Printed and bound by Printing Express, Hong Kong

You might also like to visit
www.harpercollins.co.uk
The book lover's website

# HOW THIS BOOK WILL HELP YOU

We have planned this book to make your revision as easy and effective as possible.

Here's how:

## SHORT, ACCESSIBLE NOTES THAT YOU CAN INTEGRATE INTO YOUR REVISION FILE

*Collins Revision Notes A2 Physics* has been prepared by a top examiner who knows exactly what you need to revise in order to be successful.

You can *either* base your revision on this book *or* you can tear off the notes and integrate them into your own revision file. This will ensure that you have the best possible notes to revise from.

## STUDENT-FRIENDLY PRESENTATION

The notes use visual aids – diagrams, tables, charts, etc. – so the content is easier to remember.

There is also systematic use of colour to help you revise:

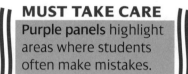

**MUST REMEMBER**
Red panels reinforce key points.

**MUST TAKE CARE**
Purple panels highlight areas where students often make mistakes.

**MUST KNOW**
Green panels highlight vital content.

– **Red type** identifies key physics terms.
– Green type identifies key definitions and equations.
– Yellow highlight emphasises important words and phrases.

## CONTENT MATCHED TO YOUR SPECIFICATION

Remember that Knowledge and Understanding of AS topics are also tested in A2 examinations.

The Contents/Specification Matching Grid on pages iv–v lists each short topic and shows which specifications it is relevant to. This means you know exactly which topics you need to revise. In some topics, there are short sections that are only relevant to one or two specifications. These are clearly marked.

All topics in the A2 core that are common to **all** specifications are included. It has not, however, been possible to include all the extension material for all specifications, in particular the Edexcel Salters-Horners and OCR B (Advancing Physics) specifications.

## GUIDANCE ON EXAM TECHNIQUE

This book concentrates on providing you with the best possible revision notes.

Worked examples are included to help you with answering exam questions. If you want more help with exam technique, then use the exam practice book alongside these Revision Notes: *Collins Do Brilliantly A2 Physics*.

Using both these books will help you to achieve a high grade in your A2 Physics exams.

# CONTENTS

| Unit | Pages | Topic | AQA(A) | AQA(B) | Edexcel | Edexcel S-H | OCR(A) | OCR (B) Ad Phys | CCEA | WJEC |
|---|---|---|---|---|---|---|---|---|---|---|
| 35 | 91-92 | Demonstrating electromagnetic induction | ✓ | ✓ | ✓ | ✓ | ✓ | ✓ | ✓ | ✓ |
| 36 | 93 | Magnitude of induced e.m.f.s | ✓ | ✓ | ✓ | ✓ | ✓ | ✓ | ✓ | ✓ |
| 37 | 94 | Eddy currents | ✓ | ✓ | ✓ | ✗ | ✓ | ✓ | ✓ | ✓ |
| 38 | 95-96 | Using electromagnetic induction | ✓ | ✓ | ✓ | ✓ | ✓ | ✓ | ✓ | ✓ |
| 39 | 97-98 | Nuclear structure | ✓ | ✗ | ✗ | ✓ | OC4 | ✓ | ✓ | ✓ |
| 40 | 99 | Nuclear density and stability | ✓ | ✗ | ✗ | ✗ | ✓ | ✓ | ✓ | ✗ |
| 41 | 100 | Radioactive decay | ✓ | ✓ | ✓ | ✓ | ✓ | ✓ | ✓ | ✓ |
| 42 | 101-102 | Properties of α, β and γ radiation | ✓ | ✗ | ✗ | ✓ | OC4 | ✓ | ✓ | ✓ |
| 43 | 103 | Experiments with radioactivity | ✓ | ✗ | ✗ | ✗ | ✓ | ✗ | ✓ | ✓ |
| 44 | 104-106 | Decay constant and half-life | ✓ | ✓ | ✓ | ✓ | ✓ | ✗ | ✓ | ✓ |
| 45 | 107-108 | Mass–energy equivalence | ✓ | ✓ | ✓ | ✓ | OC4 | ✗ | ✓ | ✓ |
| 46 | 109-111 | Nuclear fission and fusion | ✓ | ✓ | ✓ | ✓ | ✓ | ✗ | ✓ | ✓ |
| 47 | 112 | Using radioactivity | ✓ | ✓ | ✗ | ✗ | OC4 | ✓ | ✓ | ✓ |
| 48 | 113-114 | Particle physics | ✗ | ✗ | ✗ | ✗ | ✗ | ✗ | ✗ | ✗ |
| 49 | 115 | Second law of thermodynamics | OM6C | ✗ | ✗ | ✗ | ✗ | ✗ | ✓ | ✗ |
| 50 | 116 | Engine cycles | OM6C | ✗ | ✗ | ✗ | ✗ | ✗ | ✗ | ✗ |
| 51 | 117-118 | Rotational dynamics | OM6C | ✗ | ✗ | ✗ | ✗ | ✗ | ✗ | ✗ |
| 52 | 119-120 | Lenses | OM6A&6B | ✗ | ✗ | ✗ | OC2 | ✗ | ✗ | ✗ |
| 53 | 121-122 | The eye | OM6B | ✗ | ✗ | ✗ | OC2 | ✗ | ✗ | ✗ |
| 54 | 123 | The ear | OM6B | ✗ | ✗ | ✗ | OC2 | ✗ | ✗ | ✗ |
| 55 | 124 | X-rays | OM6B | ✗ | ✗ | ✗ | OC2 | ✓ | ✗ | ✗ |
| 56 | 125 | Imaging with light | OM6B | ✗ | ✗ | ✗ | ✗ | ✗ | ✗ | ✗ |
| 57 | 126-128 | Stellar evolution | OM6A | ✗ | ✗ | ✗ | OC1 | a little | ✗ | ✗ |
| 58 | 129-130 | Classification of stars | OM6A | ✗ | ✗ | ✗ | OC1 | a little | ✗ | ✗ |
| 59 | 131-134 | Telescopes | OM6A | ✗ | ✗ | ✗ | ✗ | ✗ | ✗ | ✗ |
| 60 | 135-136 | Special relativity | OM6D | ✗ | ✗ | ✗ | OC1 | ✗ | ✗ | ✗ |

✓ indicates that a significant proportion of this topic is relevant to this specification.

✗ indicates that this topic is not included in this specification.

○ indicates that this topic is relevant to an option in this specification.

Optional topics for AQA(A) and OCR(A) have been included as follows:

**AQA Specification A**

| Optional Module 6A | Astrophysics | Units 9, 52 and 57-59 |
|---|---|---|
| Optional Module 6B | Medical Physics | Units 52-56 |
| Optional Module 6C | Applied Physics | Units 49-51 |
| Optional Module 6D | Turning Points in Physics | Units 10, 25, 30 and 60 |

**OCR Specification A**

| Optional Component | OC1 Cosmology | Units 9, 21 and 57, 58 and 60 |
|---|---|---|
| Optional Component | OC2 Health Physics | Units 52-55 |
| Optional Component | OC3 Materials | Units 18 and 33 |
| Optional Component | OC4 Nuclear and Particle Physics | Units 24, 25, 31, 40, 46 and 48 |

(Electronics and Telecommunications options are not included.)

# CIRCULAR MOTION

- A body moving at constant speed in a circular path is accelerating.

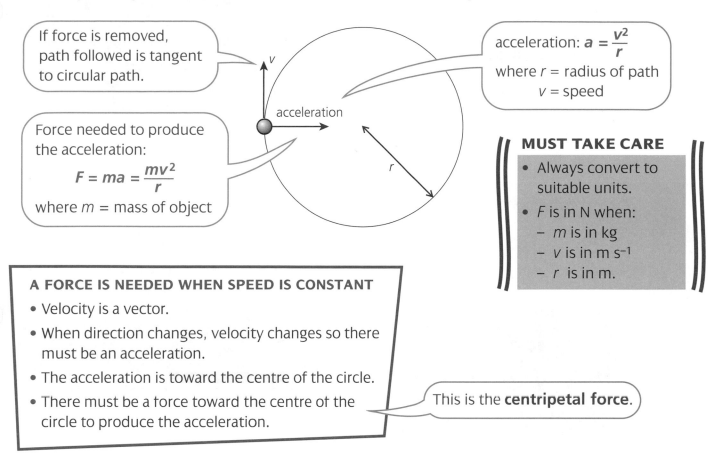

If force is removed, path followed is tangent to circular path.

Force needed to produce the acceleration:
$$F = ma = \frac{mv^2}{r}$$
where $m$ = mass of object

acceleration: $a = \dfrac{v^2}{r}$

where $r$ = radius of path
$v$ = speed

**MUST TAKE CARE**

- Always convert to suitable units.
- $F$ is in N when:
  - $m$ is in kg
  - $v$ is in m s$^{-1}$
  - $r$ is in m.

**A FORCE IS NEEDED WHEN SPEED IS CONSTANT**

- Velocity is a vector.
- When direction changes, velocity changes so there must be an acceleration.
- The acceleration is toward the centre of the circle.
- There must be a force toward the centre of the circle to produce the acceleration.

This is the **centripetal force**.

## HOW ANGULAR SPEED IS USED

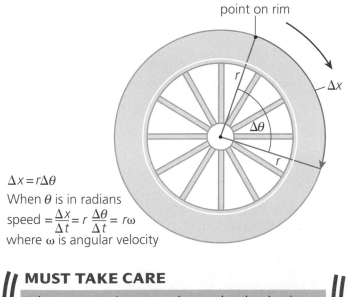

point on rim

$\Delta x = r\Delta\theta$
When $\theta$ is in radians
speed $= \dfrac{\Delta x}{\Delta t} = r\dfrac{\Delta\theta}{\Delta t} = r\omega$
where $\omega$ is angular velocity

**MUST REMEMBER**

- Angular speed, $v = r\omega$
  so $a = r\omega^2$
  $F = mr\omega^2$
- $F$ is in N when:
  - $m$ is in kg
  - $r$ is in m
  - $\omega$ is in rad s$^{-1}$
- $\omega = 2\pi f$
  when frequency $f$ is in revolutions per second
- $\omega = \dfrac{2\pi f}{60}$
  when $f$ is in revolutions per minute

**MUST TAKE CARE**

When answering questions, check whether given:

- the angular speed in radians per second
- the linear speed in metres per second
- the frequency of rotation in revolutions per second or revolutions per minute.

# HOW CENTRIPETAL FORCE OCCURS IN PRACTICE

**Car going round corner**

Centripetal force is the **frictional force** between tyres and road surface.

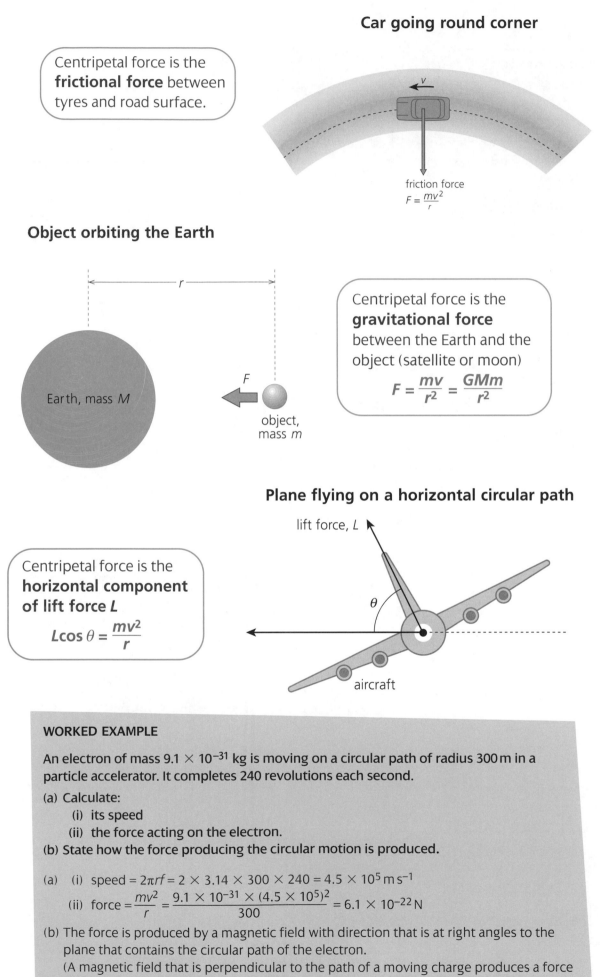

friction force
$$F = \frac{mv^2}{r}$$

## Object orbiting the Earth

Earth, mass $M$

object, mass $m$

Centripetal force is the **gravitational force** between the Earth and the object (satellite or moon)

$$F = \frac{mv}{r^2} = \frac{GMm}{r^2}$$

## Plane flying on a horizontal circular path

lift force, $L$

Centripetal force is the **horizontal component of lift force $L$**

$$L\cos\theta = \frac{mv^2}{r}$$

aircraft

**WORKED EXAMPLE**

An electron of mass $9.1 \times 10^{-31}$ kg is moving on a circular path of radius 300 m in a particle accelerator. It completes 240 revolutions each second.

(a) Calculate:
  (i) its speed
  (ii) the force acting on the electron.
(b) State how the force producing the circular motion is produced.

(a)  (i)  speed $= 2\pi r f = 2 \times 3.14 \times 300 \times 240 = 4.5 \times 10^5 \,\mathrm{m\,s^{-1}}$
   (ii)  force $= \dfrac{mv^2}{r} = \dfrac{9.1 \times 10^{-31} \times (4.5 \times 10^5)^2}{300} = 6.1 \times 10^{-22}\,\mathrm{N}$

(b) The force is produced by a magnetic field with direction that is at right angles to the plane that contains the circular path of the electron.
   (A magnetic field that is perpendicular to the path of a moving charge produces a force that is at right angles to both the field and the direction of motion of the charge.)

# WORK, ENERGY AND POWER

## WORK

- Work is only done when the point at which a force is applied moves in the direction of the force.
- When a force does work, energy is transferred from one form into another.
- **Work done = force × distance moved** in the **direction of the force**
  $$W = Fd$$
- The unit of work is the joule, J.
- 1 J is the work done when 1 N moves its point of application by 1 m.
  **1 J = 1 N m**

> In circular motion, centripetal force is at right angles to the direction of movement, so this force does no work.

> A satellite in orbit meets some resistance by colliding with gas and other solid particles, so work is done on it which slows it down and raises its temperature.

> - The load gains **potential energy** as it is lifted.
> - Gain in P.E. = $Fh$
> - If $F$ just lifts the load of mass $m$, $\Delta$**P.E. = $mgh$**

### MUST REMEMBER

- If there is friction, e.g. at pulley, some energy is used doing work against the frictional force.
- This energy appears as internal energy in the pulley and raises its temperature.

## ENERGY

- A system with energy is capable of doing work.
- Whatever produces the force to do work loses energy.
- The unit of energy is the joule, J.

> To do this work:
> - an electric motor would use **electrical energy ($E = VIt$)**
> - a person would need to convert **chemical energy**.

### MUST TAKE CARE

- Must not confuse power and force.

## POWER

- Power is the rate at which work is done or at which energy is transferred.
- If the load moves the height $h$ in time $t$,
  power, $P = \dfrac{Fh}{t}$
- $\dfrac{h}{t} = v$ so **$P = Fv$**
  where $v$ is the velocity of the load during lifting
- The unit of power is the watt, W. **1 W = 1 J s$^{-1}$**

> Work done by accelerating force = $Fx$
> Engine uses fuel to do the work.

> Useful energy gained is **kinetic energy** of the car.
> **K.E. = $\frac{1}{2}mv^2$** (= $Fx$ if no energy is lost)

> - If accelerating force is constant, the power developed increases as velocity increases.
> - Mean useful power developed during
>   time $t = \dfrac{\text{K.E. gained}}{t}$
>   $= F \times$ average velocity
>   $= F \times \dfrac{v}{2}$

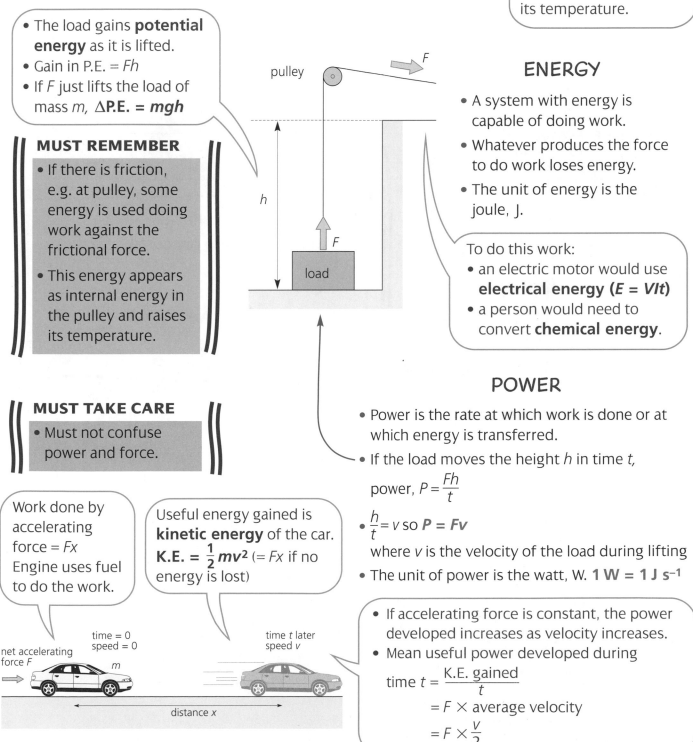

pulley

$F$

$h$

$F$

load

net accelerating force $F$

time = 0
speed = 0

$m$

time $t$ later
speed $v$

distance $x$

# SIMPLE HARMONIC MOTION (SHM)

## WHAT IS SHM?

- SHM describes oscillations about a fixed point.
- The fixed point is the centre or **equilibrium position** of the oscillation.

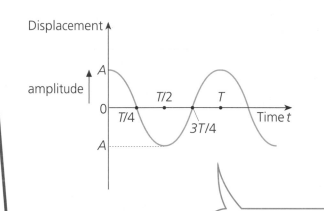

## WHAT CONDITIONS PRODUCE SHM?

- Acceleration:
  - is always toward a fixed point
  - is proportional to the displacement from the fixed point
  - is given by $a = -\omega^2 x$
    where $a$ is the acceleration,
    $x$ is the displacement and
    $\omega$ is a constant $= 2\pi f \ (= \frac{2\pi}{T})$

    Negative sign shows that acceleration is in the opposite direction to the displacement.

If displacement is a maximum when $t = 0$, then the displacement–time graph is a cosine graph.

## MUST REMEMBER

- **Displacement:**
  - is a vector
  - defines the distance and direction from the equilibrium position
  - may be + or −
- **Amplitude $A$:**
  - is the **maximum displacement** of the oscillating object from the equilibrium position.
- **Period $T$:**
  - the time for **one complete to and fro** oscillation of the motion.
- **Frequency $f$:**
  - is the number of oscillations per second.

  $f = \dfrac{1}{T}$

## EQUATION DESCRIBING SHM

- Displacement $x = A \cos (2\pi ft)$
  assuming $x = A$ when $t = 0$

  Note: $2\pi ft$ gives the angle in radians.

  Multiply by $\dfrac{180}{\pi}$ to give the angle in degrees.

- For displacement $x$, velocity $v = \pm 2\pi f \sqrt{A^2 - x^2}$

## MUST TAKE CARE

Must not confuse displacement and amplitude

## WORKED EXAMPLE

A body performs simple harmonic motion of amplitude 0.040 m with a period of 1.5 s.
(a) Write down the equation that relates the displacement $x$ and time $t$ for the motion.
(b) How far from the equilibrium position will the object be 0.60 s after reaching its maximum displacement?

(a) $x = 0.040 \cos (\frac{2\pi}{1.5} t) = 0.040 \cos 4.2t$

(b) Displacement $x = 0.040 \cos (4.2 \times 0.60 \times \frac{180}{\pi})°$

$= 0.040 \cos 144° = 0.040 \times (-0.81) = -0.032 \, \text{m}$

Since the start position is +0.040 m, the object displacement will be 0.032 m on the opposite side of the equilibrium position.

# REPRESENTING SHM USING GRAPHS

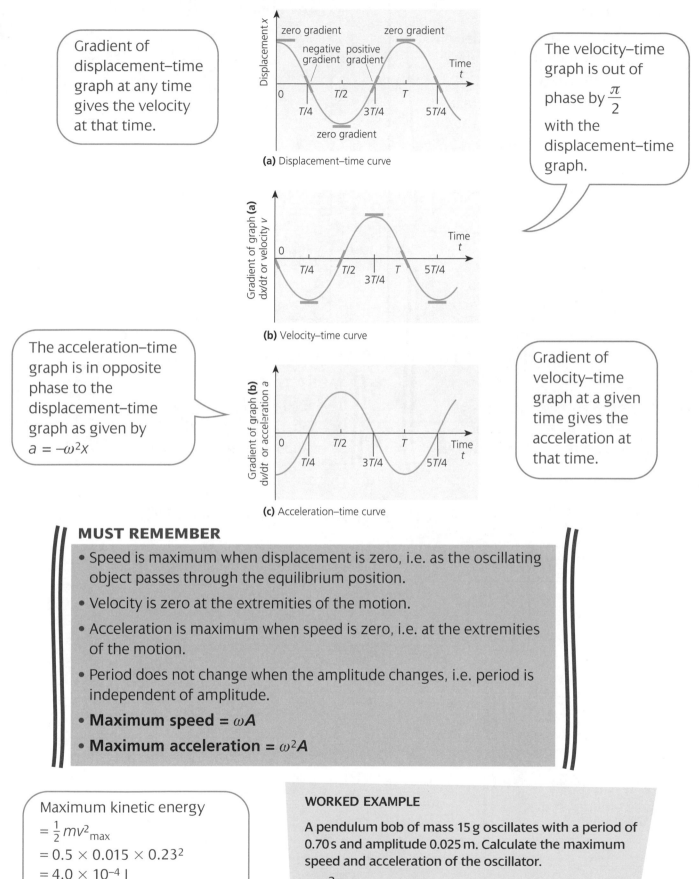

Gradient of displacement–time graph at any time gives the velocity at that time.

The velocity–time graph is out of phase by $\frac{\pi}{2}$ with the displacement–time graph.

The acceleration–time graph is in opposite phase to the displacement–time graph as given by $a = -\omega^2 x$

Gradient of velocity–time graph at a given time gives the acceleration at that time.

**(a)** Displacement–time curve

**(b)** Velocity–time curve

**(c)** Acceleration–time curve

## MUST REMEMBER

- Speed is maximum when displacement is zero, i.e. as the oscillating object passes through the equilibrium position.
- Velocity is zero at the extremities of the motion.
- Acceleration is maximum when speed is zero, i.e. at the extremities of the motion.
- Period does not change when the amplitude changes, i.e. period is independent of amplitude.
- **Maximum speed = $\omega A$**
- **Maximum acceleration = $\omega^2 A$**

Maximum kinetic energy
$= \frac{1}{2}mv^2_{max}$
$= 0.5 \times 0.015 \times 0.23^2$
$= 4.0 \times 10^{-4}$ J
This is also the total energy.

Since $F = ma$, maximum force on the bob $= 0.015 \times 2.0 = 0.030$ N

### WORKED EXAMPLE

A pendulum bob of mass 15 g oscillates with a period of 0.70 s and amplitude 0.025 m. Calculate the maximum speed and acceleration of the oscillator.

$\omega = \frac{2\pi}{T} = 9.0$ rad s$^{-1}$
maximum speed $= \omega A = 9.0 \times 0.025$
$= 0.23$ m s$^{-1}$
maximum acceleration $= \omega^2 A = 9.0^2 \times 0.025$
$= 2.0$ m s$^{-2}$

# ENERGY AND SHM

## ENERGY CHANGES

- As a mass oscillates, potential energy is transferred to kinetic energy and vice versa.
- The potential energy $E_P$ is greatest when the oscillating mass is at maximum displacement.
- The kinetic energy $E_K$ is maximum when it is moving fastest, i.e. at the centre of the oscillation.

## MUST REMEMBER

- Energy is always positive.
- K.E. is maximum twice during each cycle.
- K.E. and P.E. graphs are antiphase.
- Energy must be conserved so total energy is constant unless there is damping of oscillations.

## HOW ENERGY CHANGES WITH TIME

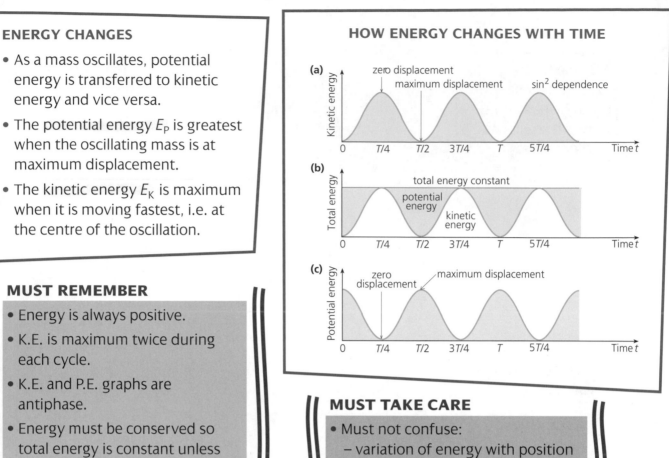

## MUST TAKE CARE

- Must not confuse:
  - variation of energy with position
  - variation of energy with time.

## HOW ENERGY CHANGES WITH POSITION

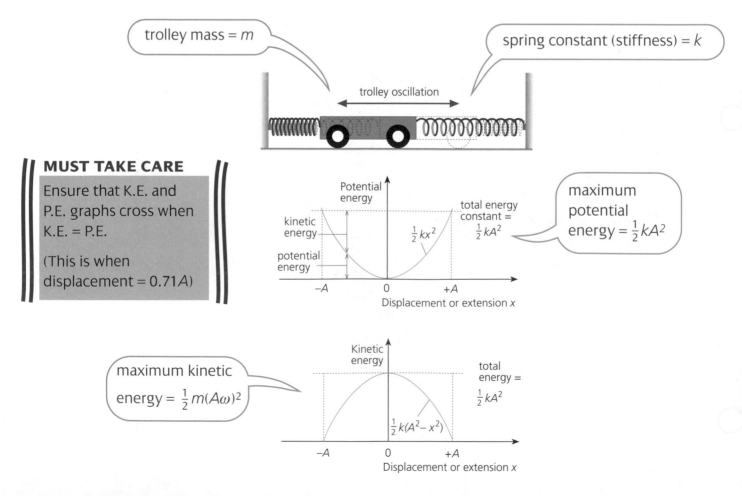

trolley mass = $m$

spring constant (stiffness) = $k$

trolley oscillation

### MUST TAKE CARE

Ensure that K.E. and P.E. graphs cross when K.E. = P.E.

(This is when displacement = 0.71$A$)

maximum potential energy = $\frac{1}{2}kA^2$

maximum kinetic energy = $\frac{1}{2}m(A\omega)^2$

# PRACTICAL SIMPLE HARMONIC OSCILLATORS

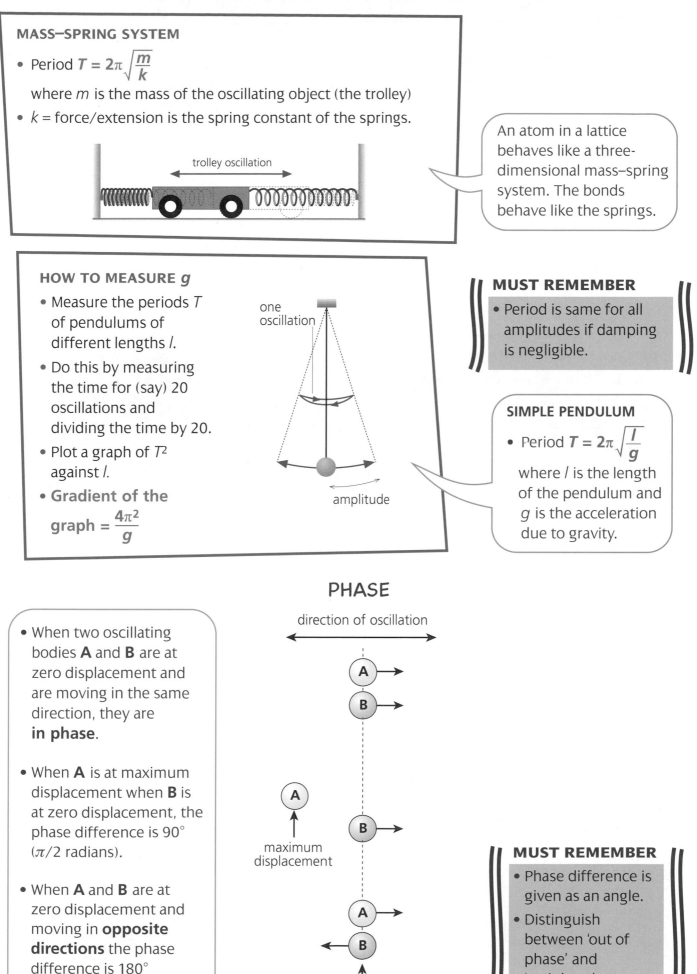

## MASS–SPRING SYSTEM

- Period $T = 2\pi\sqrt{\dfrac{m}{k}}$

  where $m$ is the mass of the oscillating object (the trolley)
- $k$ = force/extension is the spring constant of the springs.

trolley oscillation

An atom in a lattice behaves like a three-dimensional mass–spring system. The bonds behave like the springs.

## HOW TO MEASURE g

- Measure the periods $T$ of pendulums of different lengths $l$.
- Do this by measuring the time for (say) 20 oscillations and dividing the time by 20.
- Plot a graph of $T^2$ against $l$.
- **Gradient of the graph** $= \dfrac{4\pi^2}{g}$

one oscillation

amplitude

**MUST REMEMBER**

- Period is same for all amplitudes if damping is negligible.

**SIMPLE PENDULUM**

- Period $T = 2\pi\sqrt{\dfrac{l}{g}}$

  where $l$ is the length of the pendulum and $g$ is the acceleration due to gravity.

## PHASE

- When two oscillating bodies **A** and **B** are at zero displacement and are moving in the same direction, they are **in phase**.

- When **A** is at maximum displacement when **B** is at zero displacement, the phase difference is 90° ($\pi/2$ radians).

- When **A** and **B** are at zero displacement and moving in **opposite directions** the phase difference is 180° ($\pi$ radians, **antiphase**).

direction of oscillation

A

B

A

B

maximum displacement

A

B

centre of oscillation

**MUST REMEMBER**

- Phase difference is given as an angle.
- Distinguish between 'out of phase' and 'antiphase'.

# FREE AND FORCED OSCILLATIONS AND RESONANCE

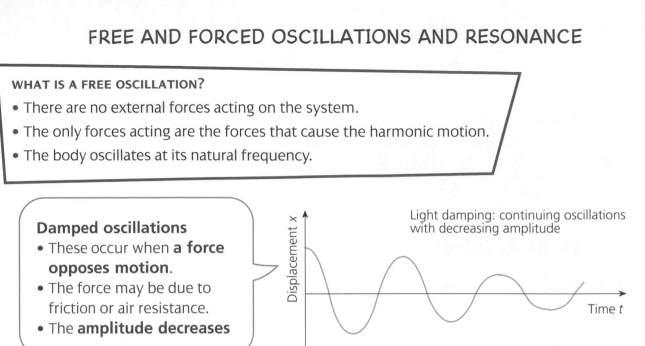

**WHAT IS A FREE OSCILLATION?**
- There are no external forces acting on the system.
- The only forces acting are the forces that cause the harmonic motion.
- The body oscillates at its natural frequency.

**Damped oscillations**
- These occur when **a force opposes motion**.
- The force may be due to friction or air resistance.
- The **amplitude decreases**

Light damping: continuing oscillations with decreasing amplitude

**WHAT IS A FORCED OSCILLATION?**
- This occurs when an external periodic force acts on an oscillator.
- It oscillates at the frequency of the applied force.
- Amplitude builds up until the energy put in each second by the external periodic force is equal to that lost due to damping.

**MUST REMEMBER**
- Heavier damping makes the amplitude fall more quickly.
- Heavier damping means more resistive force (more friction).

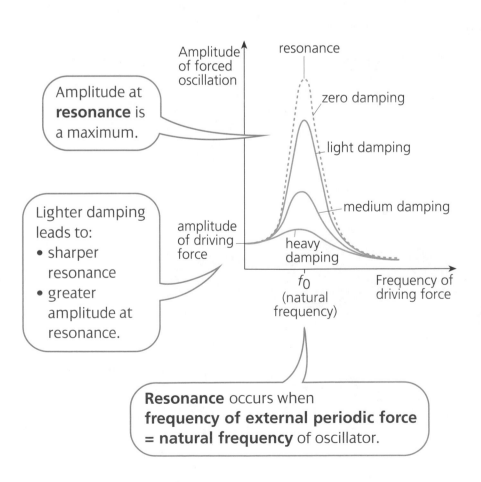

Amplitude at **resonance** is a maximum.

Lighter damping leads to:
- sharper resonance
- greater amplitude at resonance.

**Resonance** occurs when **frequency of external periodic force = natural frequency** of oscillator.

**EXAMPLES OF RESONANCE AND FORCED OSCILLATIONS**
- Vibrations produced in machines – resonance occurs when motor frequency matches natural frequency of a component, e.g. car wing mirror
- Use of resonance in electrical circuits to tune in to a radio station – resonant frequency is frequency of transmission
- An atom trapped in a lattice has a natural frequency and will readily absorb electromagnetic waves with that frequency.

# WAVES

## HOW WAVES ARE FORMED

- A source produces the disturbance which can be:
  - a pulse or continuous vibration
  - a sudden single movement, e.g. an explosion, that produces a pulse.
- A vibrating source, e.g. a vibrating string, produces a continuous wave.

### TRAVELLING (PROGRESSIVE) WAVE

- Source gives energy to particles of the transmitting medium near to it.
- Particles oscillate with the same frequency as the source.
- Energy is passed to the next particles in the transmitting medium.
- Position of maximum displacement moves along the medium.
- Each particle oscillates slightly out of phase with the one before it.

### MUST REMEMBER

- Particles stay in the same mean position.
- Only energy moves along the path of the wave.

### TRANSVERSE WAVES

Particles in the transmitting medium **oscillate at right angles** to the direction in which the **wave energy travels**.

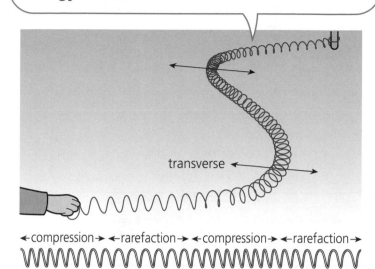

←compression→ ←rarefaction→ ←compression→ ←rarefaction→

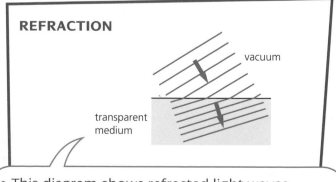

### LONGITUDINAL WAVES (e.g. SOUND)

- Particles (coils of spring) **oscillate along the direction** in which the **wave energy travels**.
- The spring is compressed and energy is then passed to the next section of the spring.
- In sound:
  - a **compression** is a region of high pressure
  - a **rarefaction** is a region of low pressure.

## WAVE PROPERTIES

### REFLECTION

incident wave    reflected wave

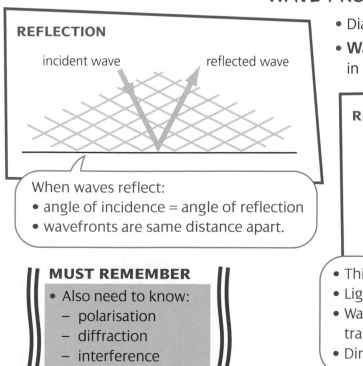

When waves reflect:
- angle of incidence = angle of reflection
- wavefronts are same distance apart.

### MUST REMEMBER

- Also need to know:
  - polarisation
  - diffraction
  - interference

- Diagrams show peak wavefronts.
- **Wavefronts** are adjacent points that oscillate in phase.

### REFRACTION

vacuum

transparent medium

- This diagram shows refracted light waves.
- Light travels slower in the transparent medium.
- Wavefronts are closer together in the transparent medium.
- Direction of travel changes towards the normal.

# WAVE VELOCITY

- **Wave velocity** is:
  - the velocity (speed) at which energy travels
  - the velocity at which a peak travels.
- Sound travels:
  - at about $340\,\mathrm{m\,s^{-1}}$ in air at room temperature
  - faster in water and faster still in solids.

- Need good contact between transducers and the surface.
- In medicine, a gel is used to ensure good transmission of energy.

### Ultrasound
- These are frequencies above 20 000 Hz.
- Higher frequencies give better definition and detect smaller flaws.
- Ultrasound is used:
  - in medical diagnosis
  - to detect flaws in metals.

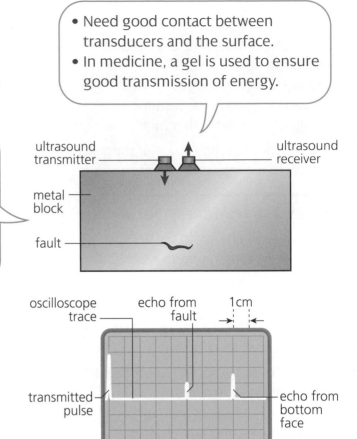

ultrasound transmitter — ultrasound receiver

metal block

fault

oscilloscope trace — echo from fault — 1cm

transmitted pulse — echo from bottom face

time base set to $5\,\mu\mathrm{s\,cm^{-1}}$

**WORKED EXAMPLE**

From the diagrams calculate:
(a) the thickness of the block
(b) the position of the fault.
Speed of ultrasound in the metal is $5100\,\mathrm{m\,s^{-1}}$.

(a) Time for pulse to go from top to bottom and back $= 8 \times 5\,\mu\mathrm{s} = 40\,\mu\mathrm{s}$
Distance travelled by the pulse
$= vt$
$= 40 \times 10^{-6} \times 5100\,\mathrm{m}$
$= 0.204\,\mathrm{m}$ (to 3 s.f.)
Thickness of block $= 0.102\,\mathrm{m}$
(b) The pulse takes $25\,\mu\mathrm{s}$ to travel to the fault and back.
Depth of fault below top surface
$= \frac{1}{2}(25 \times 10^{-6} \times 5100)\,\mathrm{m}$
$= 0.064\,\mathrm{m}$

**MUST REMEMBER**

- A pulse has to travel **to and from** the reflector.
- Distance travelled by wave is 2 × distance from source to reflector.

## WAVE VELOCITY = FREQUENCY x WAVELENGTH

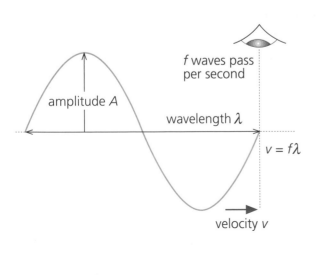

amplitude $A$

wavelength $\lambda$

$f$ waves pass per second

$v = f\lambda$

velocity $v$

- $v = f\lambda$
- If $f$ waves pass a point each second, then in $1/f$ seconds one wave passes the point and wave travels a distance $\lambda$.
- Wave speed, $v = \dfrac{\lambda}{1/f} = f\lambda$

**MUST REMEMBER**

- **Wavelength,** $\lambda$ is:
  - the distance between two successive peaks of the wave
  - the minimum distance between two points in the medium that are oscillating in phase.

# ELECTROMAGNETIC WAVES

## ELECTROMAGNETIC (E-M) WAVES

- E-m waves are transverse waves.
- They travel at $3.00 \times 10^8 \text{ m s}^{-1}$ in a vacuum.
- They travel slower in other media:
  - light travels at about $2 \times 10^8 \text{ m s}^{-1}$ in glass
  - radio waves travel at about $2 \times 10^8 \text{ m s}^{-1}$ through polythene in coaxial cable.

**MUST KNOW**
- Approximate wavelengths for each region

Visible light, infrared and ultraviolet radiation result from electrons falling between an atom's higher energy levels.

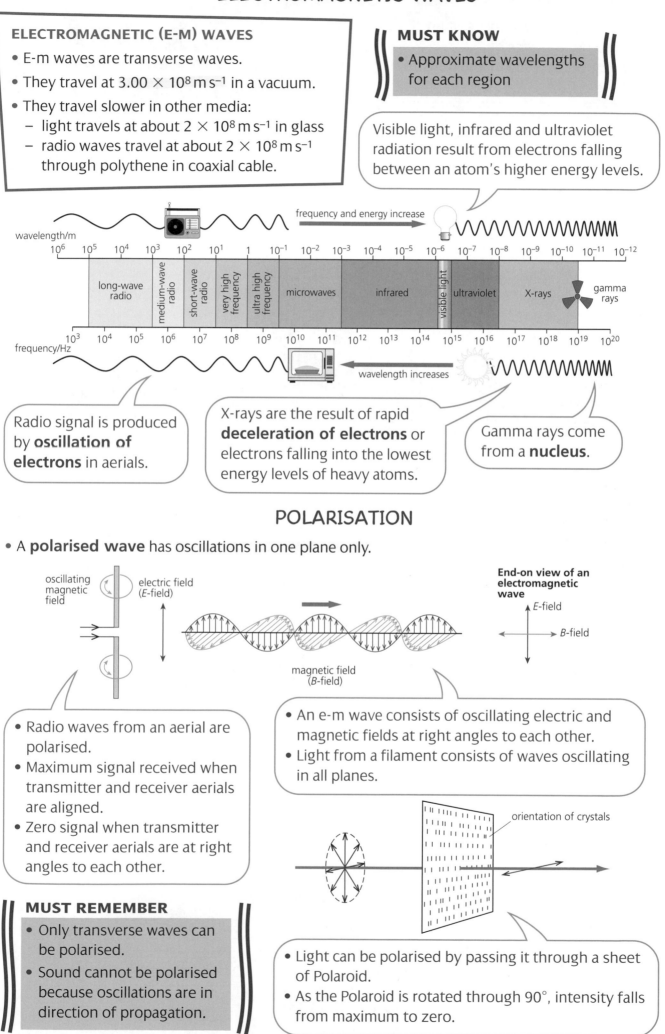

Radio signal is produced by **oscillation of electrons** in aerials.

X-rays are the result of rapid **deceleration of electrons** or electrons falling into the lowest energy levels of heavy atoms.

Gamma rays come from a **nucleus**.

## POLARISATION

- A **polarised wave** has oscillations in one plane only.

**End-on view of an electromagnetic wave**

- Radio waves from an aerial are polarised.
- Maximum signal received when transmitter and receiver aerials are aligned.
- Zero signal when transmitter and receiver aerials are at right angles to each other.

- An e-m wave consists of oscillating electric and magnetic fields at right angles to each other.
- Light from a filament consists of waves oscillating in all planes.

**MUST REMEMBER**
- Only transverse waves can be polarised.
- Sound cannot be polarised because oscillations are in direction of propagation.

- Light can be polarised by passing it through a sheet of Polaroid.
- As the Polaroid is rotated through 90°, intensity falls from maximum to zero.

# INVERSE SQUARE LAW

- The inverse square law applies when:
  - there is a point source
  - no energy is absorbed by the transmitting medium.

> The inverse square law applies because energy is conserved.

> Applies particularly to e-m radiation: light, X-rays and gamma rays

> Astronomical distances are large compared with size of stars so stars can be treated as point sources.

## MUST REMEMBER

- Power of a source is $P$ watts.
- Intensity, or energy flux, $I$ is:
  - power per square metre
  - measured in $W\,m^{-2}$

2r

point source

> At distance $2r$ from the source, intensity =
> $$\frac{P}{4\pi(2r)^2} = \frac{1}{4}\frac{P}{4\pi r^2}$$

- Energy spreads out equally in all directions.
- At distance $r$ from the source, the energy is spread over surface of a sphere of radius $r$.
- $P$ joules per second are spread over area of $4\pi r^2\,m^2$.
- **Intensity** $= \dfrac{P}{4\pi r^2}\,J\,s^{-1}\,m^{-2}$ or $W\,m^{-2}$

> **Double** distance gives $\frac{1}{4}$ intensity, **treble** distance gives $\frac{1}{9}$ intensity, and so on.

## WORKED EXAMPLE

The intensity of radiation falling on the outer atmosphere of the Earth from the Sun is $1400\,W\,m^{-2}$. The Earth is $1.5 \times 10^{11}\,m$ from the Sun.

Calculate:

(a) the total power radiated by the Sun
(b) the intensity of radiation falling on Neptune which is $4.5 \times 10^{12}\,m$ from the Sun.

(a) Intensity $= \dfrac{P}{4\pi r^2}$

$$1400 = \frac{P}{4\pi(1.5 \times 10^{11})^2}$$

$$P = 4\pi \times 1400 \times (1.5 \times 10^{11})^2$$
$$= 4.0 \times 10^{26}\,W$$

(b) Intensity $\propto \dfrac{1}{r^2}$

$$\frac{\text{Intensity on Neptune}}{\text{Intensity on Earth}} = \frac{(\text{Distance of Earth from Sun})^2}{(\text{Distance of Neptune from Sun})^2}$$

$$\text{Intensity on Neptune} = \frac{(1.5 \times 10^{11})^2}{(4.5 \times 10^{12})^2} \times 1400 = 1.6\,W\,m^{-2}$$

## MUST TAKE CARE

- Check the ratio is the right way round.
- Check the intensity further from the source is lower.

# STATIONARY (STANDING) WAVES

## SUPERPOSITION

- Superposition produces stationary waves and interference patterns.
- When two (or more) waves meet at a point they superpose.
- Resultant displacement is the sum of the displacements of the waves at that time.

**MUST KNOW**

**Stationary waves** are produced when **two waves** of the **same frequency** travel in **opposite directions** through the **same medium.**

This sequence shows stationary wave production on a stretched string. In each diagram the incident wave moves $\lambda/8$ to the right and the reflected wave $\lambda/8$ to the left.

A **stationary wave** is produced by **superposition** of the **incident** and **reflected waves.**

**MUST TAKE CARE. . .**
- The distance between two nodes is half a wavelength.
- one wavelength = length of two loops

- A **node** is a point of **zero amplitude**.
- The displacement is **always** zero.
- Any fixed point must be a node.

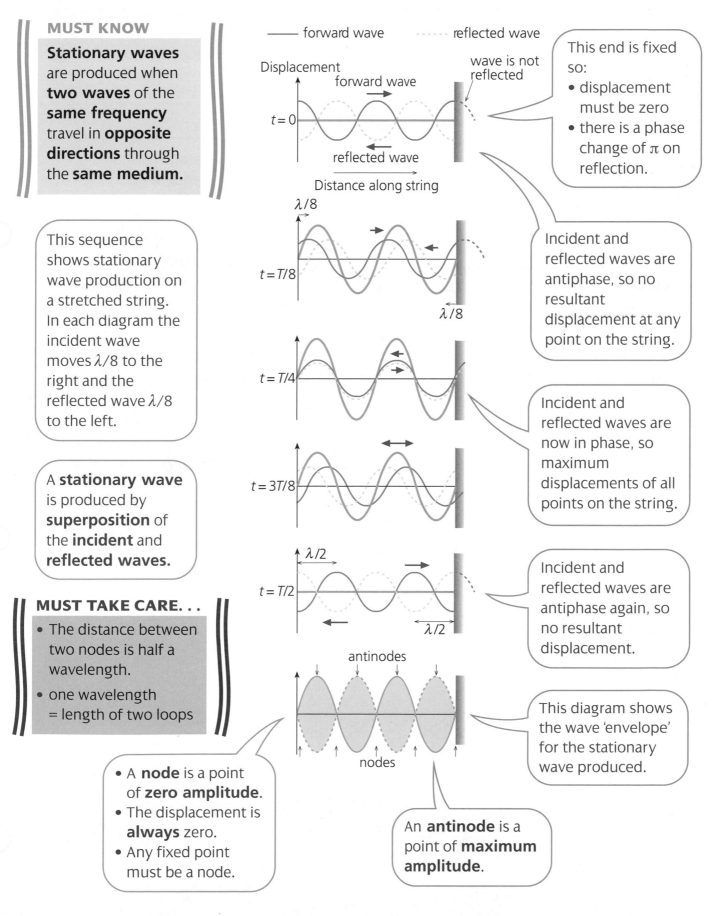

This end is fixed so:
- displacement must be zero
- there is a phase change of $\pi$ on reflection.

Incident and reflected waves are antiphase, so no resultant displacement at any point on the string.

Incident and reflected waves are now in phase, so maximum displacements of all points on the string.

Incident and reflected waves are antiphase again, so no resultant displacement.

This diagram shows the wave 'envelope' for the stationary wave produced.

An **antinode** is a point of **maximum amplitude**.

# STATIONARY WAVES ON A STRETCHED STRING

This is an example of **resonance**.
Only particular well-defined frequencies of the vibration generator produce stationary waves on the stretched string.

- **Fundamental frequency** is the lowest frequency note the string can emit.
- $f_0 = \dfrac{1}{2l}\sqrt{\dfrac{T}{\mu}}$

  where $l$ is length of the string
  $T$ is the tension
  $\mu$ is the mass per metre of the string.

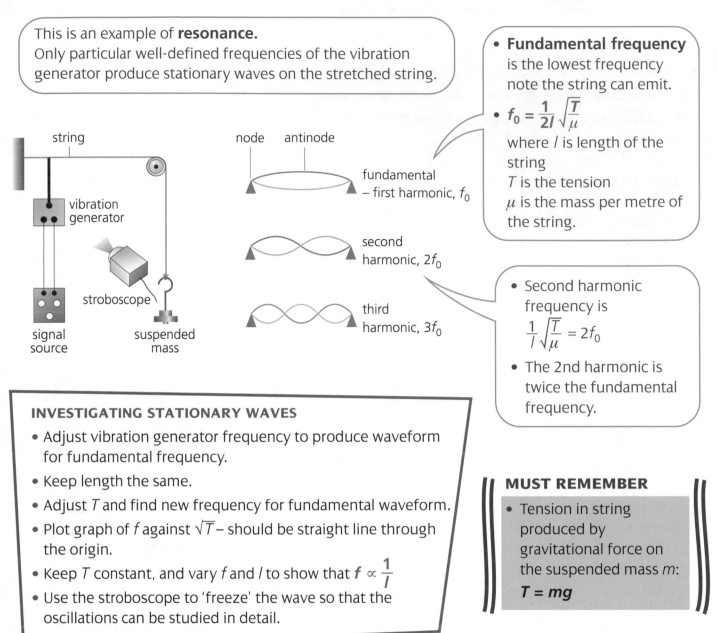

- Second harmonic frequency is

  $\dfrac{1}{l}\sqrt{\dfrac{T}{\mu}} = 2f_0$

- The 2nd harmonic is twice the fundamental frequency.

**INVESTIGATING STATIONARY WAVES**

- Adjust vibration generator frequency to produce waveform for fundamental frequency.
- Keep length the same.
- Adjust $T$ and find new frequency for fundamental waveform.
- Plot graph of $f$ against $\sqrt{T}$ – should be straight line through the origin.
- Keep $T$ constant, and vary $f$ and $l$ to show that $f \propto \dfrac{1}{l}$
- Use the stroboscope to 'freeze' the wave so that the oscillations can be studied in detail.

**MUST REMEMBER**

- Tension in string produced by gravitational force on the suspended mass $m$:
  **$T = mg$**

# STATIONARY WAVES IN PIPES

**MUST TAKE CARE**

- Open ends must be antinodes and closed ends nodes.
- Air cannot oscillate at closed end.

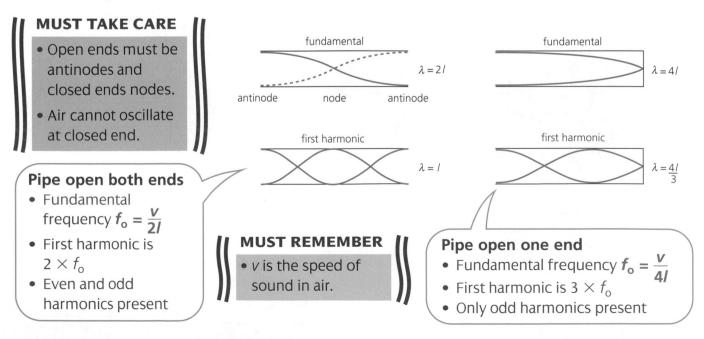

**Pipe open both ends**
- Fundamental frequency $f_o = \dfrac{v}{2l}$
- First harmonic is $2 \times f_o$
- Even and odd harmonics present

**MUST REMEMBER**

- $v$ is the speed of sound in air.

**Pipe open one end**
- Fundamental frequency $f_o = \dfrac{v}{4l}$
- First harmonic is $3 \times f_o$
- Only odd harmonics present

# DIFFRACTION

## WHAT IS DIFFRACTION?

- It is the spreading of waves when they pass through a gap or meet an obstacle in their path.
- It is a property of waves – particles would go straight on.

## MUST TAKE CARE

- For waves through a slit:
  - for small angles $\frac{\lambda}{b} = \theta$ in radians
  - for large angles $\frac{\lambda}{b} = \sin\theta$

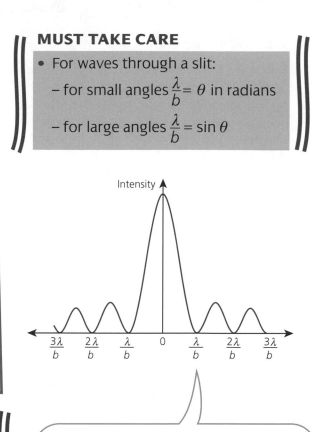

## MUST REMEMBER

- Central maximum is twice as wide and much more intense than the others.
- Most of the energy falls in the central maximum.

- If slit width $b = \lambda$, minimum is at 90°
- If slit width $b = 2\lambda$, the first minimum is at 30°

## DIFFRACTION USING AERIALS

- Microwave energy from a dish or horn aperture is diffracted at the aperture.
- For a circular dish of diameter $D$, the first minimum is at $\frac{1.22\lambda}{D}$ radians.
- $\frac{\lambda}{D}$ gives an approximate value for the angle.

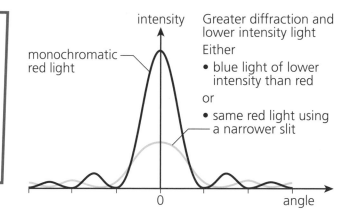

intensity

monochromatic red light

Greater diffraction and lower intensity light
Either
- blue light of lower intensity than red

or
- same red light using a narrower slit

angle

## WORKED EXAMPLE

Which of the following produces the wider central maximum when viewing a distant star: a 40 m diameter dish tuned to receive 0.1 m wavelength microwaves or a naked eye of pupil diameter 5 mm using visible light of wavelength 500 nm?

For eye image, $\theta = \dfrac{500 \times 10^{-9}}{5 \times 10^{-3}} = 1 \times 10^{-4}$ radians

For dish image, $\theta = \dfrac{0.1}{40} = 2.5 \times 10^{-3}$ radians

The dish produces the wider central maximum so the resolution would be poorer (see page 18).

## MUST REMEMBER

- Central maximum is wider for:
  - a narrower slit
  - a longer wavelength.

# IMAGES OF STARS

**WORKED EXAMPLE**

(a) The image of a star is produced by a telescope system with a circular aperture of diameter 0.20 m. Calculate the angular width of the central maximum of the image of a star in blue light of wavelength 450 nm.

(b) The image is formed 0.50 m from the focusing lens. Calculate the diameter of the central maximum of the image formed.

(a) Using diffraction formula for a circular aperture,

$\sin \theta = \dfrac{1.22\lambda}{D}$ where $D$ = diameter of aperture

$\sin \theta = \dfrac{1.22 \times 450 \times 10^{-9}}{0.20} = 2.75 \times 10^{-6}$

This is a small angle so $\theta = 2.75 \times 10^{-6}$ radians

Width of central maximum $= 2\theta$

$= 5.5 \times 10^{-6}$ radians

(b) Central maximum diameter $= 0.5 \times 5.5 \times 10^{-6}$ m

$= 2.8 \times 10^{-6}$ m $(=2.8\,\mu\text{m})$

Most energy falls in a circle of this diameter.

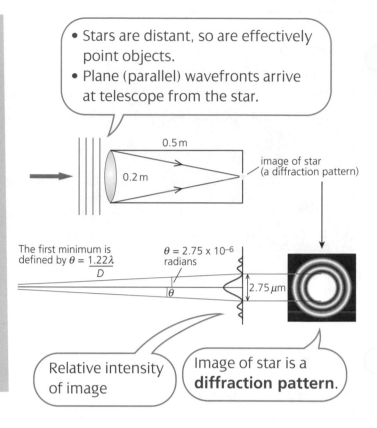

- Stars are distant, so are effectively point objects.
- Plane (parallel) wavefronts arrive at telescope from the star.

0.5 m

image of star (a diffraction pattern)

0.2 m

The first minimum is defined by $\theta = \dfrac{1.22\lambda}{D}$

$\theta = 2.75 \times 10^{-6}$ radians

$2.75\,\mu\text{m}$

Relative intensity of image

Image of star is a **diffraction pattern**.

# RESOLUTION

- Objects are resolved when they can be seen as separate objects.

- **Resolution** is defined as **the minimum angle subtended by the objects at the observing aperture that enables two objects to be seen as separate objects.**

**MUST REMEMBER**

- Sources are **just resolved** when the central maximum of one pattern coincides with the first minimum of the other.

- For better resolution (better **resolving power**):
  – view using shorter wavelength radiation
  – use wider apertures.

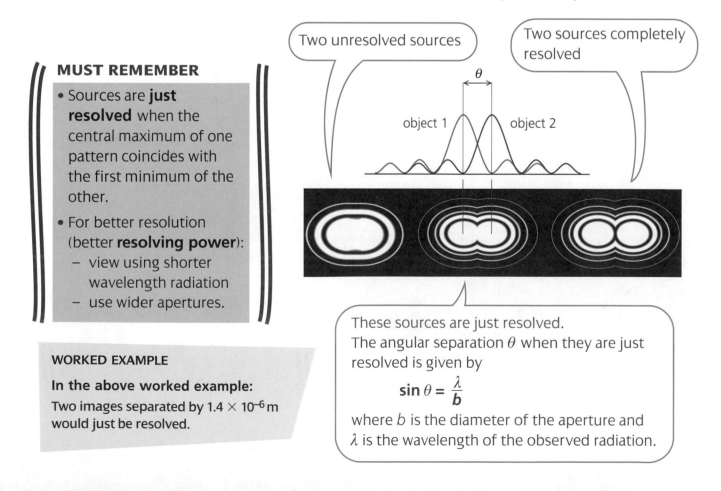

Two unresolved sources

Two sources completely resolved

$\theta$

object 1     object 2

These sources are just resolved.
The angular separation $\theta$ when they are just resolved is given by

$$\sin \theta = \frac{\lambda}{b}$$

where $b$ is the diameter of the aperture and $\lambda$ is the wavelength of the observed radiation.

**WORKED EXAMPLE**

**In the above worked example:**

Two images separated by $1.4 \times 10^{-6}$ m would just be resolved.

# INTERFERENCE

- **Interference patterns** are formed by superposition of two or more coherent waves.
- **Coherent waves** have the same frequency and constant phase difference.

## YOUNG'S TWO-SLIT EXPERIMENT

Monochromatic light from a distant source – a point source or narrow slit

**MUST KNOW**

**Monochromatic light** has a single wavelength.

Interference pattern along XY where the beams overlap

Light emerging from A and B is coherent because the waves come from the same source.

Light waves diffract at A and B producing overlapping beams.

- The **path difference** results in a **phase difference** between the waves arriving at P.
- Phase difference,
$$\Delta(\text{phase}) = \frac{2\pi\,\Delta(\text{path})}{\lambda}$$

To reach P the wave from A has to travel further than the wave from B.
The **path difference**, $\Delta(\text{path}) = AP - BP$

**MUST REMEMBER**

- Path difference = $n\lambda$
  - Waves are in phase.
  - Amplitude and intensity are maximum.
- Path difference = $n\lambda + \dfrac{\lambda}{2}$
  $$= \left(n + \tfrac{1}{2}\right)\lambda$$
  - Waves are in antiphase so they cancel each other.
  - Amplitude and intensity are minimum.

Intensity can only be zero if light from each slit has equal intensity and P is equal distance from each slit.

**MUST KNOW**

- A suitable apparatus to demonstrate Young's fringes
- Typical dimensions to produce visible fringes

about 3 metres

lamp with line filament

blackened microscope slide with two narrow lines scratched on it about 2 mm apart

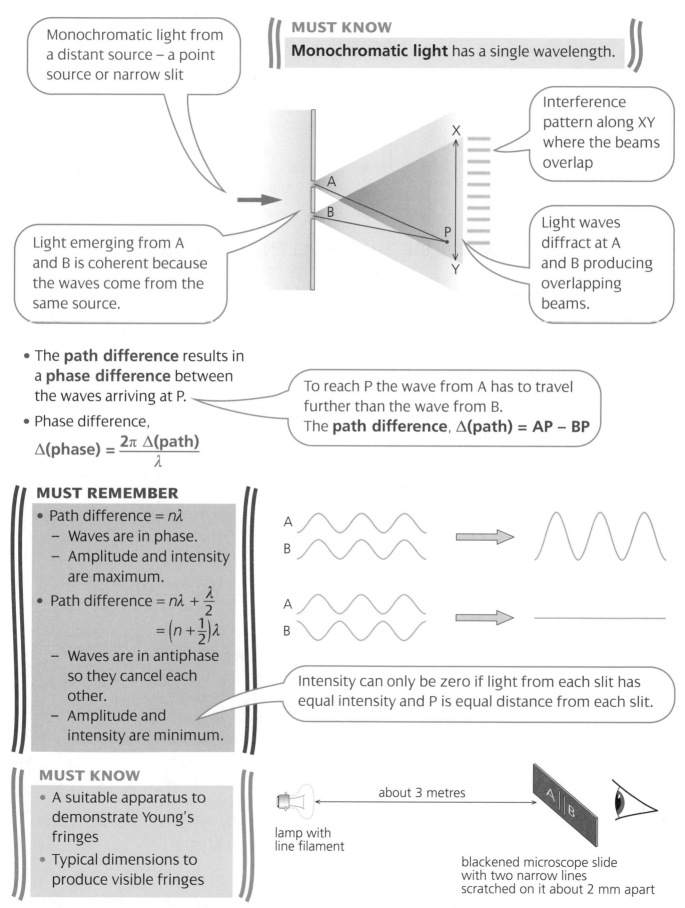

# FRINGE SPACING

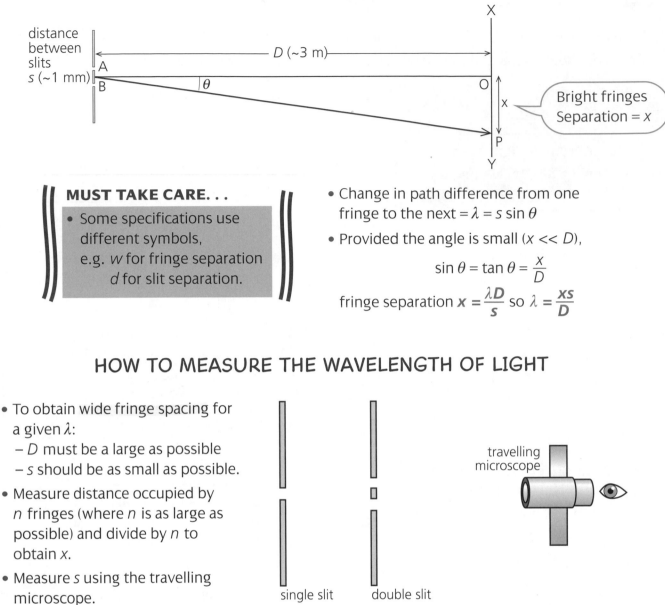

distance between slits
$s$ (~1 mm)

A

B

$\theta$

$D$ (~3 m)

X

O

x

P

Y

Bright fringes
Separation = $x$

**MUST TAKE CARE. . .**

- Some specifications use different symbols, e.g. $w$ for fringe separation $d$ for slit separation.

- Change in path difference from one fringe to the next = $\lambda = s \sin \theta$
- Provided the angle is small ($x \ll D$),

$$\sin \theta = \tan \theta = \frac{x}{D}$$

fringe separation $x = \dfrac{\lambda D}{s}$ so $\lambda = \dfrac{xs}{D}$

## HOW TO MEASURE THE WAVELENGTH OF LIGHT

- To obtain wide fringe spacing for a given $\lambda$:
  - $D$ must be a large as possible
  - $s$ should be as small as possible.
- Measure distance occupied by $n$ fringes (where $n$ is as large as possible) and divide by $n$ to obtain $x$.
- Measure $s$ using the travelling microscope.
- Measure $D$ with a metre rule.

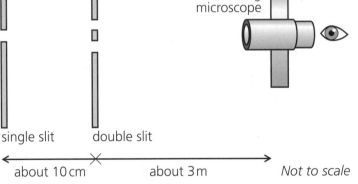

travelling microscope

single slit    double slit

about 10 cm    about 3 m    *Not to scale*

## HOW TO MEASURE THE WAVELENGTH OF MICROWAVES

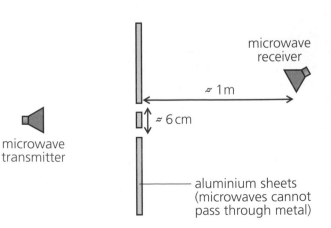

microwave receiver

≈ 1 m

≈ 6 cm

microwave transmitter

aluminium sheets (microwaves cannot pass through metal)

- Straight-through position is a maximum.
- Move receiver to find next maximum (wavelength is about 3 cm so with 6 cm slit spacing this will be at about 30°).
- If receiver is 1 m from slits:
  - measure angular separation of the maximums and slit separation
  - use $s \sin \theta = \lambda$ to find $\lambda$.
- If receiver is close to slits, find wavelength by measuring path difference directly using a metre rule.

# DIFFRACTION GRATING

- Diffraction gratings consist of many closely spaced slits.
- Lines are cut onto a piece of glass. The transparent parts are the slits.

## USING DIFFRACTION GRATINGS
- The light from many slits superposes in some places to produce maximum brightness.
- The regions of high light intensity for a given wavelength are narrow.

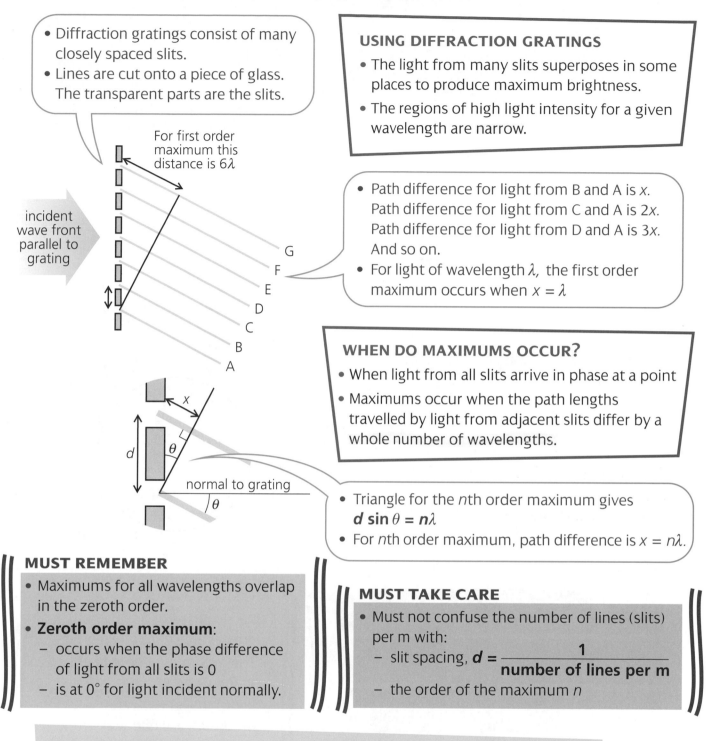

For first order maximum this distance is $6\lambda$

incident wave front parallel to grating

- Path difference for light from B and A is $x$. Path difference for light from C and A is $2x$. Path difference for light from D and A is $3x$. And so on.
- For light of wavelength $\lambda$, the first order maximum occurs when $x = \lambda$

normal to grating

## WHEN DO MAXIMUMS OCCUR?
- When light from all slits arrive in phase at a point
- Maximums occur when the path lengths travelled by light from adjacent slits differ by a whole number of wavelengths.

- Triangle for the $n$th order maximum gives
  $$d \sin \theta = n\lambda$$
- For $n$th order maximum, path difference is $x = n\lambda$.

## MUST REMEMBER
- Maximums for all wavelengths overlap in the zeroth order.
- **Zeroth order maximum**:
  - occurs when the phase difference of light from all slits is 0
  - is at 0° for light incident normally.

## MUST TAKE CARE
- Must not confuse the number of lines (slits) per m with:
  - slit spacing, $d = \dfrac{1}{\text{number of lines per m}}$
  - the order of the maximum $n$

## WORKED EXAMPLE

Light of wavelength 630 nm falls normally on a grating that has 350 lines per mm.
Calculate:
(a) the angle at which the first order maximum occurs
(b) the total number of maximums that are formed by the grating.

(a) Number of lines per m = 350 000

so $d = \dfrac{1}{350\,000} = 2.86 \times 10^{-6}$ m

For first order, $d \sin \theta = \lambda$

$\sin \theta = \dfrac{\lambda}{d} = \dfrac{630 \times 10^{-9}}{2.86 \times 10^{-6}} = 0.22$

$\theta = 12.7°$

(b) For $n$th order, $d \sin \theta = n\lambda$
For highest order visible, $\sin \theta \leq 1$

$\dfrac{n\lambda}{d} \leq 1 \Rightarrow n \leq \dfrac{d}{\lambda}$

$n \leq \dfrac{2.86 \times 10^{-6}}{630 \times 10^{-9}} \Rightarrow n \leq 4.5$

$n$ must be whole integer so highest order is the 4th.
Total maximums visible is 9
(central maximum + 4 on each side).

# ADVANTAGES OF DIFFRACTION GRATINGS

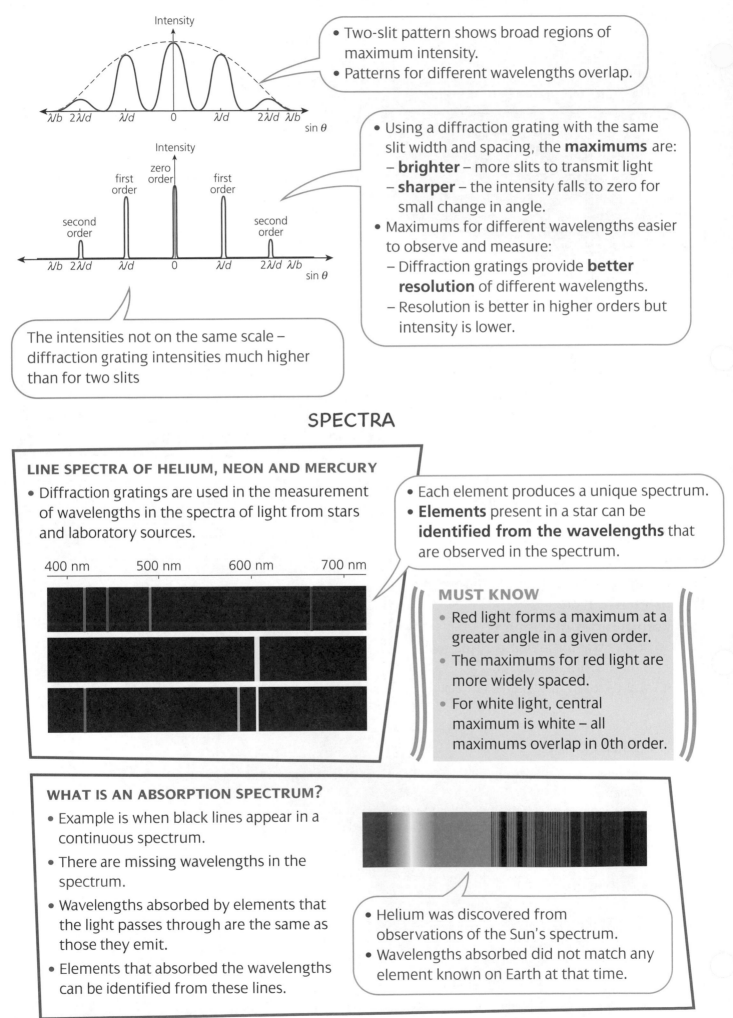

• Two-slit pattern shows broad regions of maximum intensity.
• Patterns for different wavelengths overlap.

• Using a diffraction grating with the same slit width and spacing, the **maximums** are:
  – **brighter** – more slits to transmit light
  – **sharper** – the intensity falls to zero for small change in angle.
• Maximums for different wavelengths easier to observe and measure:
  – Diffraction gratings provide **better resolution** of different wavelengths.
  – Resolution is better in higher orders but intensity is lower.

The intensities not on the same scale – diffraction grating intensities much higher than for two slits

# SPECTRA

## LINE SPECTRA OF HELIUM, NEON AND MERCURY
• Diffraction gratings are used in the measurement of wavelengths in the spectra of light from stars and laboratory sources.

400 nm     500 nm     600 nm     700 nm

• Each element produces a unique spectrum.
• **Elements** present in a star can be **identified from the wavelengths** that are observed in the spectrum.

### MUST KNOW
• Red light forms a maximum at a greater angle in a given order.
• The maximums for red light are more widely spaced.
• For white light, central maximum is white – all maximums overlap in 0th order.

## WHAT IS AN ABSORPTION SPECTRUM?
• Example is when black lines appear in a continuous spectrum.
• There are missing wavelengths in the spectrum.
• Wavelengths absorbed by elements that the light passes through are the same as those they emit.
• Elements that absorbed the wavelengths can be identified from these lines.

• Helium was discovered from observations of the Sun's spectrum.
• Wavelengths absorbed did not match any element known on Earth at that time.

# DOPPLER EFFECT

- This is the change in observed frequency or wavelength as a wave source and an observer move toward or away from one another.
- When relative motion $v$ is small compared with wave speed $c$ ($v \ll c$)

$$\frac{\text{change in frequency}}{\text{frequency from stationary source}} = \frac{\Delta f}{f} \approx \frac{\Delta \lambda}{\lambda} \approx \frac{v}{c}$$

**MUST TAKE CARE**

- $f$ and $\lambda$ are the frequency and wavelength of waves from the stationary source.

- As the source moves away from the observer:
  - observed wavelength increases
  - observed frequency decreases.

**Example from sound**
To a stationary observer, the frequency of note from the siren of a police car or ambulance is higher (higher pitch) as it approaches than when it moves away.

## RED SHIFT

- This is the shift to longer wavelengths observed when a light source is moving away from an observer on the Earth.

- These are the **recession velocities** – the velocities at which the galaxies are **moving away** from the Earth.
- The velocities can be calculated from measurements of the wavelength using $\dfrac{v}{c} = \dfrac{\Delta \lambda}{\lambda}$

Spectrum from source in a laboratory

- Spectra from galaxies moving away from the Earth
- The H and K lines of the spectrum from the galaxies have a longer wavelength.

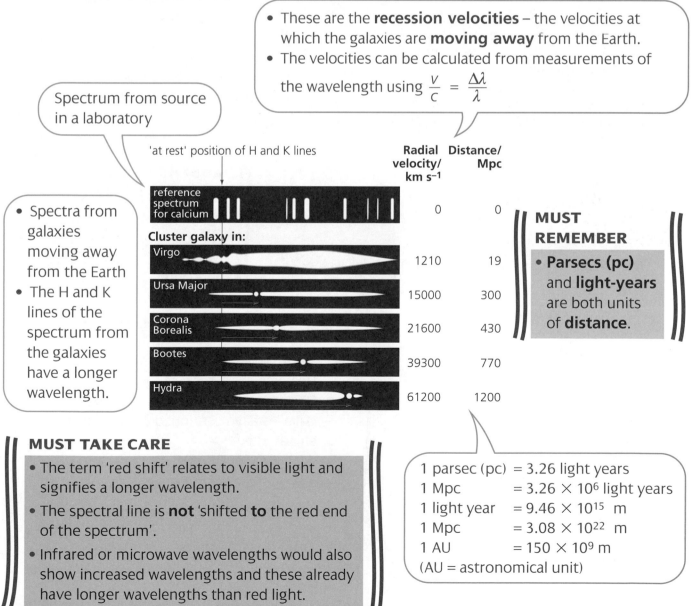

| | Radial velocity/ km s$^{-1}$ | Distance/ Mpc |
|---|---|---|
| reference spectrum for calcium | 0 | 0 |
| **Cluster galaxy in:** | | |
| Virgo | 1210 | 19 |
| Ursa Major | 15000 | 300 |
| Corona Borealis | 21600 | 430 |
| Bootes | 39300 | 770 |
| Hydra | 61200 | 1200 |

'at rest' position of H and K lines

**MUST REMEMBER**

- **Parsecs (pc)** and **light-years** are both units of **distance**.

**MUST TAKE CARE**

- The term 'red shift' relates to visible light and signifies a longer wavelength.
- The spectral line is **not** 'shifted **to** the red end of the spectrum'.
- Infrared or microwave wavelengths would also show increased wavelengths and these already have longer wavelengths than red light.

1 parsec (pc) = 3.26 light years
1 Mpc = 3.26 × 10$^6$ light years
1 light year = 9.46 × 10$^{15}$ m
1 Mpc = 3.08 × 10$^{22}$ m
1 AU = 150 × 10$^9$ m
(AU = astronomical unit)

# HUBBLE'S LAW

- This is a law based on astronomical measurements.
- It relates recession velocities of galaxies to their distances from the Earth.

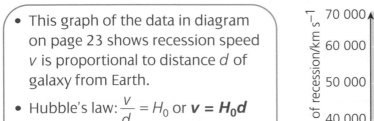

- This graph of the data in diagram on page 23 shows recession speed $v$ is proportional to distance $d$ of galaxy from Earth.
- Hubble's law: $\dfrac{v}{d} = H_0$ or $\boldsymbol{v = H_0 d}$
  where $H_0$ is the **Hubble constant**.

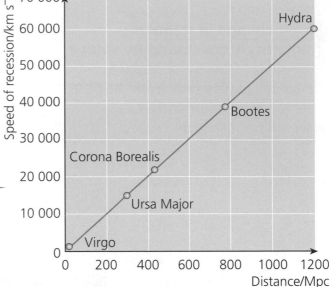

**WORKED EXAMPLE**

A wavelength in the calcium spectrum, measured using a laboratory source, is 422.4 nm. The same line in the light from Hydra has a wavelength of 508.6 nm.
(a) Calculate the recession speed of Hydra relative to Earth.
(b) The distance of Hydra from Earth is $3.9 \times 10^9$ light years. Calculate a value for the Hubble constant in $km\,s^{-1}\,Mpc^{-1}$. Speed of e-m radiation $= 3.0 \times 10^8\,m\,s^{-1}$

$$\frac{v}{c} = \frac{\Delta\lambda}{\lambda}$$

(a) $\dfrac{v}{3.0 \times 10^8} = \dfrac{(508.6 - 422.4) \times 10^{-9}}{422.4 \times 10^{-9}} = 0.204$

$v = 0.204 \times 3.0 \times 10^8 = 6.1 \times 10^7\,m\,s^{-1}$

Velocity is away from Earth since wavelength is longer.

(b) $v = 6.1 \times 10^4\,km\,s^{-1}$

$d = \dfrac{3.9 \times 10^9}{3.26 \times 10^6} = 1200\,Mpc$

$H_0 = \dfrac{v}{d} = 61\,000/1200 = 51\,km\,s^{-1}\,Mpc^{-1}$

Distances and velocities are difficult to measure accurately, so value of $H_0$ is uncertain. Accepted value is between 50 and 100 $km\,s^{-1}\,Mpc^{-1}$ (between $1.6 \times 10^{-18}\,s^{-1}$ and $3.2 \times 10^{-18}\,s^{-1}$)

**MUST TAKE CARE**

- Questions may be set in a mix of units, so may need to convert before substituting in equations.

These data suggest that the age of the universe is

$\dfrac{1}{51} \approx 0.020\,Mpc\,km^{-1}\,s$

1 Mpc $= 3.1 \times 10^{19}\,km$ so age is about $0.020 \times 3.1 \times 10^{19}\,s$
$= 6.1 \times 10^{17}\,s$ or $1.9 \times 10^{10}$ years

**HOW OLD IS THE UNIVERSE?**

- Assumptions:
  - Universe has a definite age.
  - Universe is the result of a 'big bang'.
  - Galaxies have continued to move apart at steady speeds.
- A galaxy moving with speed $v$ has moved distance $d$ since the Universe was formed.

  Time taken, $T = \dfrac{d}{v} = \dfrac{1}{H_0}$ = age of Universe

- The Universe may have expanded quicker in the early stages so calculated age may be too high.

# WAVE–PARTICLE DUALITY

- **Diffraction** and **interference** phenomena suggest that light is a wave.
  - Waves deliver energy as a continuous stream.
  - Energy is always arriving at a surface on which a wave falls.

- **Photoelectric effect** suggests that light energy is delivered in the same way as a stream of particles.
  - Energy arrives in small 'packets' called **quanta** (singular quantum).
  - A quantum of electromagnetic radiation is called a **photon**.

- Electromagnetic radiation sometimes behaves as a wave and sometimes as a stream of particles:
  - To find out where the energy goes, treat as a wave.
  - To find out what happens when energy is transferred, treat as a particle.

- The electromagnetic wave behaviour is said to show **wave–particle duality**.

## PHOTOELECTRIC EFFECT

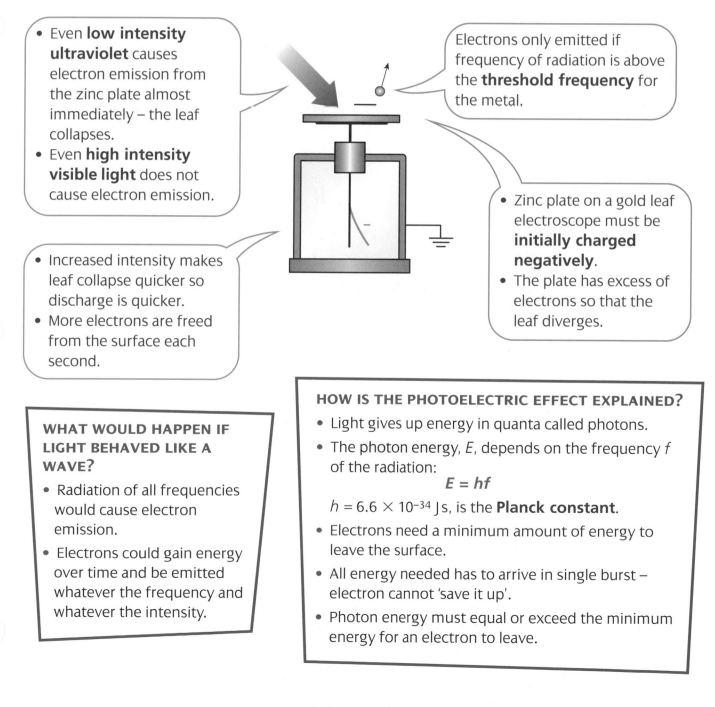

- Even **low intensity ultraviolet** causes electron emission from the zinc plate almost immediately – the leaf collapses.
- Even **high intensity visible light** does not cause electron emission.

Electrons only emitted if frequency of radiation is above the **threshold frequency** for the metal.

- Zinc plate on a gold leaf electroscope must be **initially charged negatively**.
- The plate has excess of electrons so that the leaf diverges.

- Increased intensity makes leaf collapse quicker so discharge is quicker.
- More electrons are freed from the surface each second.

**WHAT WOULD HAPPEN IF LIGHT BEHAVED LIKE A WAVE?**

- Radiation of all frequencies would cause electron emission.
- Electrons could gain energy over time and be emitted whatever the frequency and whatever the intensity.

**HOW IS THE PHOTOELECTRIC EFFECT EXPLAINED?**

- Light gives up energy in quanta called photons.
- The photon energy, $E$, depends on the frequency $f$ of the radiation:

$$E = hf$$

$h = 6.6 \times 10^{-34}$ Js, is the **Planck constant**.

- Electrons need a minimum amount of energy to leave the surface.
- All energy needed has to arrive in single burst – electron cannot 'save it up'.
- Photon energy must equal or exceed the minimum energy for an electron to leave.

# EINSTEIN PHOTOELECTRIC EQUATION

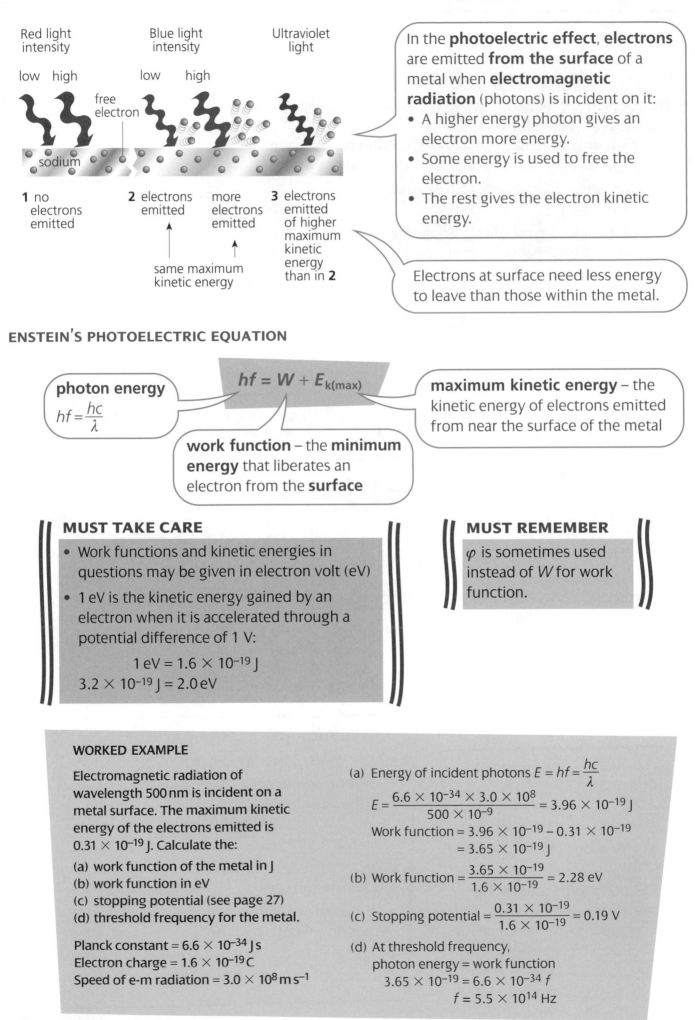

Red light intensity
low  high

Blue light intensity
low  high

Ultraviolet light

free electron

sodium

**1** no electrons emitted

**2** electrons emitted

more electrons emitted

**3** electrons emitted of higher maximum kinetic energy than in **2**

same maximum kinetic energy

In the **photoelectric effect**, **electrons** are emitted **from the surface** of a metal when **electromagnetic radiation** (photons) is incident on it:
- A higher energy photon gives an electron more energy.
- Some energy is used to free the electron.
- The rest gives the electron kinetic energy.

Electrons at surface need less energy to leave than those within the metal.

## ENSTEIN'S PHOTOELECTRIC EQUATION

$$hf = W + E_{k(max)}$$

**photon energy**
$$hf = \frac{hc}{\lambda}$$

**maximum kinetic energy** – the kinetic energy of electrons emitted from near the surface of the metal

**work function** – the **minimum energy** that liberates an electron from the **surface**

### MUST TAKE CARE

- Work functions and kinetic energies in questions may be given in electron volt (eV)
- 1 eV is the kinetic energy gained by an electron when it is accelerated through a potential difference of 1 V:

$$1\,eV = 1.6 \times 10^{-19}\,J$$
$$3.2 \times 10^{-19}\,J = 2.0\,eV$$

### MUST REMEMBER

$\varphi$ is sometimes used instead of $W$ for work function.

### WORKED EXAMPLE

Electromagnetic radiation of wavelength 500 nm is incident on a metal surface. The maximum kinetic energy of the electrons emitted is $0.31 \times 10^{-19}$ J. Calculate the:

(a) work function of the metal in J
(b) work function in eV
(c) stopping potential (see page 27)
(d) threshold frequency for the metal.

Planck constant = $6.6 \times 10^{-34}$ J s
Electron charge = $1.6 \times 10^{-19}$ C
Speed of e-m radiation = $3.0 \times 10^8$ m s$^{-1}$

(a) Energy of incident photons $E = hf = \dfrac{hc}{\lambda}$

$$E = \frac{6.6 \times 10^{-34} \times 3.0 \times 10^8}{500 \times 10^{-9}} = 3.96 \times 10^{-19}\,J$$

Work function = $3.96 \times 10^{-19} - 0.31 \times 10^{-19}$
$$= 3.65 \times 10^{-19}\,J$$

(b) Work function = $\dfrac{3.65 \times 10^{-19}}{1.6 \times 10^{-19}} = 2.28$ eV

(c) Stopping potential = $\dfrac{0.31 \times 10^{-19}}{1.6 \times 10^{-19}} = 0.19$ V

(d) At threshold frequency,
photon energy = work function
$$3.65 \times 10^{-19} = 6.6 \times 10^{-34}\,f$$
$$f = 5.5 \times 10^{14}\,Hz$$

# INVESTIGATING THE PHOTOLECTRIC EFFECT

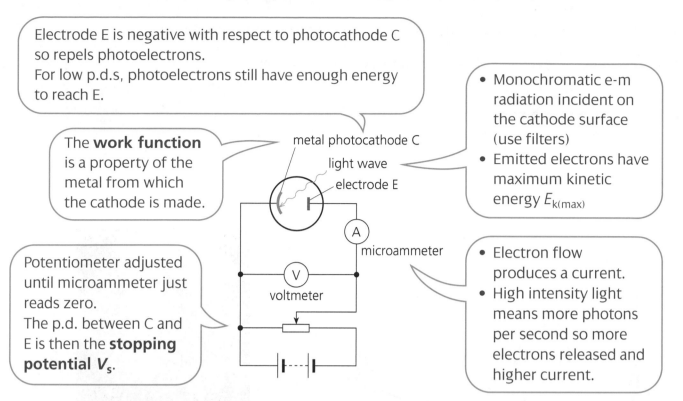

Electrode E is negative with respect to photocathode C so repels photoelectrons.
For low p.d.s, photoelectrons still have enough energy to reach E.

The **work function** is a property of the metal from which the cathode is made.

Potentiometer adjusted until microammeter just reads zero.
The p.d. between C and E is then the **stopping potential $V_s$**.

metal photocathode C

light wave

electrode E

A microammeter

V voltmeter

- Monochromatic e-m radiation incident on the cathode surface (use filters)
- Emitted electrons have maximum kinetic energy $E_{k(max)}$

- Electron flow produces a current.
- High intensity light means more photons per second so more electrons released and higher current.

- In an investigation, values of $V_s$ can be measured for different wavelengths of incident radiation and corresponding $E_{k(max)}$ calculated.

## HOW $E_{k(max)}$ IS MEASURED

- When current just reaches zero, the voltage just stops electrons with maximum energy crossing the gap.
- Energy needed for electrons to cross the gap when voltage is $V_s$ is $eV_s$:
  $E_{k(max)} = eV_s$ where $e$ is the electron charge.
- $V_s$ is the stopping potential.
- Values of $V_s$ can be measured for different wavelengths of incident radiation.

## GRAPH FROM PHOTOELECTRIC EFFECT EXPERIMENT

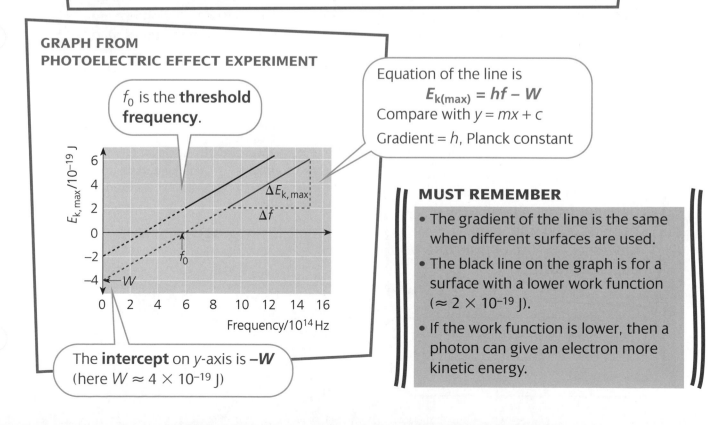

$f_0$ is the **threshold frequency**.

Equation of the line is
$$E_{k(max)} = hf - W$$
Compare with $y = mx + c$
Gradient = $h$, Planck constant

$\Delta E_{k, max}$

$\Delta f$

### MUST REMEMBER

- The gradient of the line is the same when different surfaces are used.
- The black line on the graph is for a surface with a lower work function ($\approx 2 \times 10^{-19}$ J).
- If the work function is lower, then a photon can give an electron more kinetic energy.

The **intercept** on $y$-axis is $-W$
(here $W \approx 4 \times 10^{-19}$ J)

# PARTICLES AND WAVES

- Louis de Broglie proposed that particles should exhibit **wave–particle duality**.

- The wavelength is given by the **de Broglie equation**: $\lambda = \dfrac{h}{p}$ or $\lambda = \dfrac{h}{mv}$

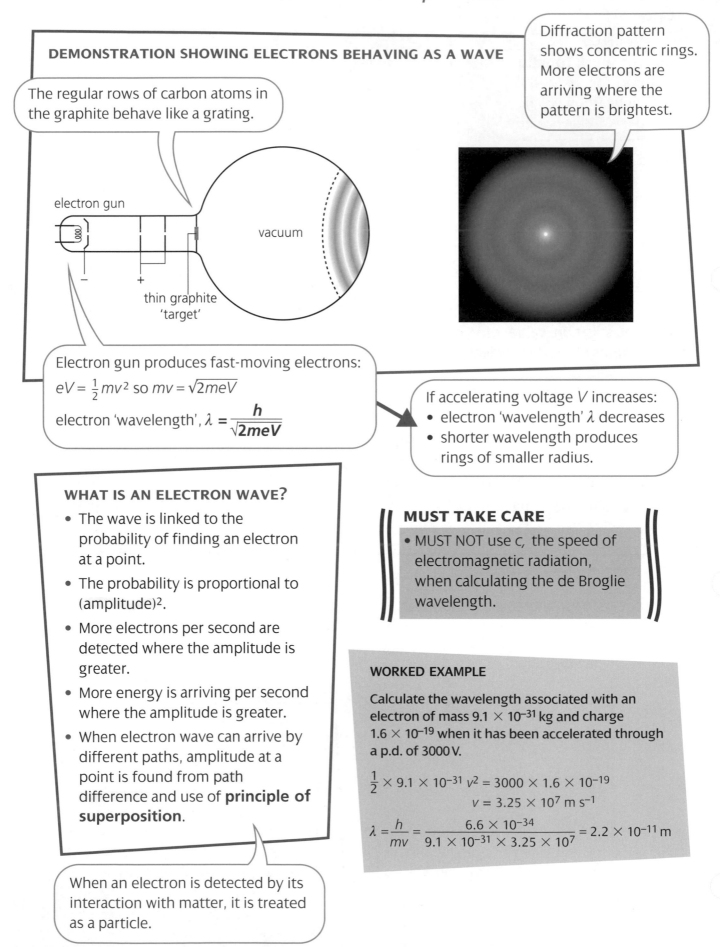

**DEMONSTRATION SHOWING ELECTRONS BEHAVING AS A WAVE**

Diffraction pattern shows concentric rings. More electrons are arriving where the pattern is brightest.

The regular rows of carbon atoms in the graphite behave like a grating.

electron gun

vacuum

thin graphite 'target'

Electron gun produces fast-moving electrons:

$eV = \frac{1}{2}mv^2$ so $mv = \sqrt{2meV}$

electron 'wavelength', $\lambda = \dfrac{h}{\sqrt{2meV}}$

If accelerating voltage $V$ increases:
- electron 'wavelength' $\lambda$ decreases
- shorter wavelength produces rings of smaller radius.

## WHAT IS AN ELECTRON WAVE?

- The wave is linked to the probability of finding an electron at a point.

- The probability is proportional to (amplitude)².

- More electrons per second are detected where the amplitude is greater.

- More energy is arriving per second where the amplitude is greater.

- When electron wave can arrive by different paths, amplitude at a point is found from path difference and use of **principle of superposition**.

When an electron is detected by its interaction with matter, it is treated as a particle.

## MUST TAKE CARE

- MUST NOT use $c$, the speed of electromagnetic radiation, when calculating the de Broglie wavelength.

**WORKED EXAMPLE**

Calculate the wavelength associated with an electron of mass $9.1 \times 10^{-31}$ kg and charge $1.6 \times 10^{-19}$ when it has been accelerated through a p.d. of 3000 V.

$$\frac{1}{2} \times 9.1 \times 10^{-31}\, v^2 = 3000 \times 1.6 \times 10^{-19}$$

$$v = 3.25 \times 10^7 \text{ m s}^{-1}$$

$$\lambda = \frac{h}{mv} = \frac{6.6 \times 10^{-34}}{9.1 \times 10^{-31} \times 3.25 \times 10^7} = 2.2 \times 10^{-11}\text{ m}$$

# ELECTRONS IN ATOMS

- De Broglie's idea of electron waves can be used to explain some facts about nuclear and atomic structure.

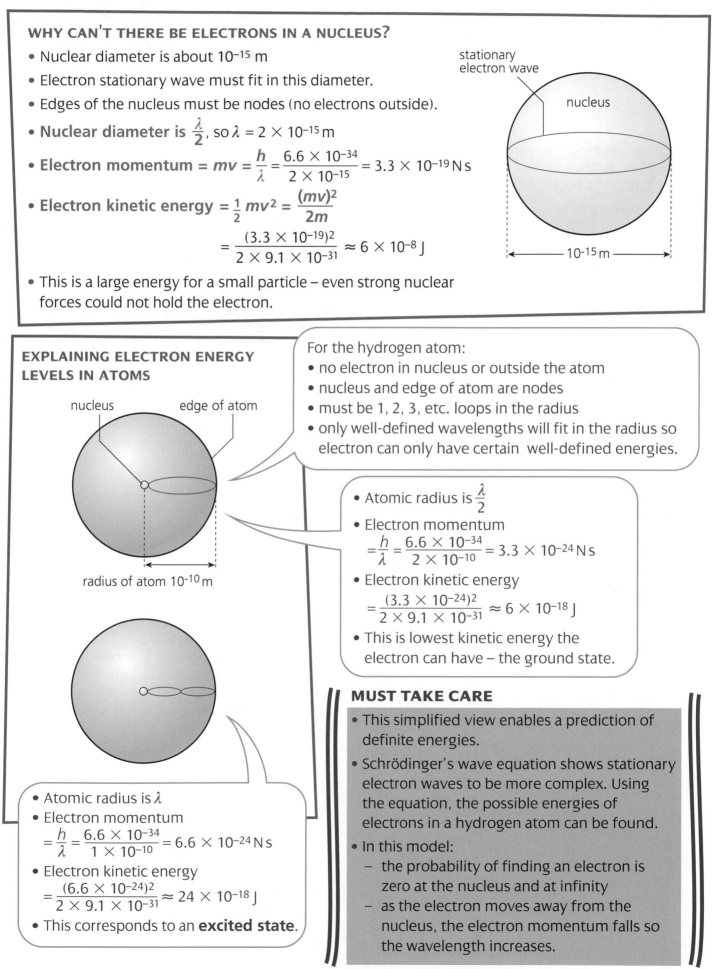

## WHY CAN'T THERE BE ELECTRONS IN A NUCLEUS?

- Nuclear diameter is about $10^{-15}$ m
- Electron stationary wave must fit in this diameter.
- Edges of the nucleus must be nodes (no electrons outside).
- **Nuclear diameter is $\frac{\lambda}{2}$**, so $\lambda = 2 \times 10^{-15}$ m
- **Electron momentum** $= mv = \dfrac{h}{\lambda} = \dfrac{6.6 \times 10^{-34}}{2 \times 10^{-15}} = 3.3 \times 10^{-19}$ N s
- **Electron kinetic energy** $= \frac{1}{2} mv^2 = \dfrac{(mv)^2}{2m}$

$$= \dfrac{(3.3 \times 10^{-19})^2}{2 \times 9.1 \times 10^{-31}} \approx 6 \times 10^{-8} \text{ J}$$

- This is a large energy for a small particle – even strong nuclear forces could not hold the electron.

stationary electron wave

nucleus

$10^{-15}$ m

## EXPLAINING ELECTRON ENERGY LEVELS IN ATOMS

nucleus            edge of atom

radius of atom $10^{-10}$ m

For the hydrogen atom:
- no electron in nucleus or outside the atom
- nucleus and edge of atom are nodes
- must be 1, 2, 3, etc. loops in the radius
- only well-defined wavelengths will fit in the radius so electron can only have certain well-defined energies.

- Atomic radius is $\frac{\lambda}{2}$
- Electron momentum
  $= \dfrac{h}{\lambda} = \dfrac{6.6 \times 10^{-34}}{2 \times 10^{-10}} = 3.3 \times 10^{-24}$ N s
- Electron kinetic energy
  $= \dfrac{(3.3 \times 10^{-24})^2}{2 \times 9.1 \times 10^{-31}} \approx 6 \times 10^{-18}$ J
- This is lowest kinetic energy the electron can have – the ground state.

- Atomic radius is $\lambda$
- Electron momentum
  $= \dfrac{h}{\lambda} = \dfrac{6.6 \times 10^{-34}}{1 \times 10^{-10}} = 6.6 \times 10^{-24}$ N s
- Electron kinetic energy
  $= \dfrac{(6.6 \times 10^{-24})^2}{2 \times 9.1 \times 10^{-31}} \approx 24 \times 10^{-18}$ J
- This corresponds to an **excited state**.

## MUST TAKE CARE

- This simplified view enables a prediction of definite energies.
- Schrödinger's wave equation shows stationary electron waves to be more complex. Using the equation, the possible energies of electrons in a hydrogen atom can be found.
- In this model:
  - the probability of finding an electron is zero at the nucleus and at infinity
  - as the electron moves away from the nucleus, the electron momentum falls so the wavelength increases.

# ENERGY LEVELS

- These are the only energies that electrons can have when in an atom.
- The atoms of an element have a unique set of energy levels.

## ENERGY LEVELS OF HYDROGEN

- An electron in an energy level higher than the ground state is in an excited state.
- Energy has to be provided to promote electrons into exited states.
- The energy can be provided by heating or by passing a current through a gas.
- Electrons can only stay in excited states for a short time.

An electron that is **just free** of the atom is defined to have **zero energy**.

The **ground state** is the lowest energy state that an electron can have. At low temperatures electrons are in low energy states.

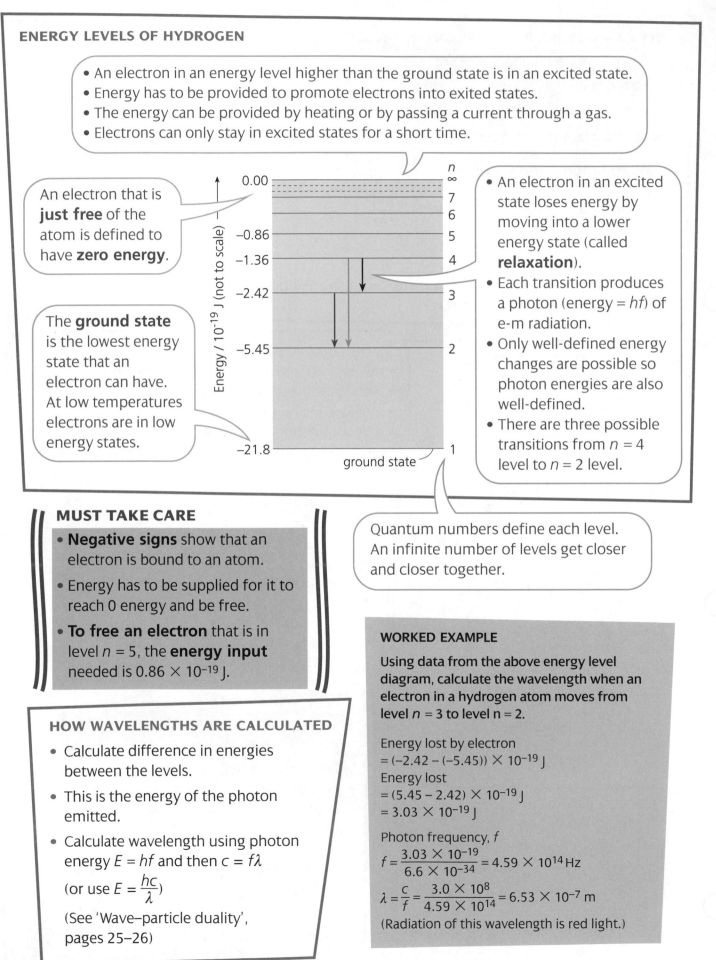

- An electron in an excited state loses energy by moving into a lower energy state (called **relaxation**).
- Each transition produces a photon (energy = $hf$) of e-m radiation.
- Only well-defined energy changes are possible so photon energies are also well-defined.
- There are three possible transitions from $n = 4$ level to $n = 2$ level.

Quantum numbers define each level. An infinite number of levels get closer and closer together.

## MUST TAKE CARE

- **Negative signs** show that an electron is bound to an atom.
- Energy has to be supplied for it to reach 0 energy and be free.
- **To free an electron** that is in level $n = 5$, the **energy input** needed is $0.86 \times 10^{-19}$ J.

## HOW WAVELENGTHS ARE CALCULATED

- Calculate difference in energies between the levels.
- This is the energy of the photon emitted.
- Calculate wavelength using photon energy $E = hf$ and then $c = f\lambda$
  (or use $E = \dfrac{hc}{\lambda}$)
  (See 'Wave–particle duality', pages 25–26)

### WORKED EXAMPLE

Using data from the above energy level diagram, calculate the wavelength when an electron in a hydrogen atom moves from level $n = 3$ to level n = 2.

Energy lost by electron
$= (-2.42 - (-5.45)) \times 10^{-19}$ J
Energy lost
$= (5.45 - 2.42) \times 10^{-19}$ J
$= 3.03 \times 10^{-19}$ J

Photon frequency, $f$
$f = \dfrac{3.03 \times 10^{-19}}{6.6 \times 10^{-34}} = 4.59 \times 10^{14}$ Hz
$\lambda = \dfrac{c}{f} = \dfrac{3.0 \times 10^{8}}{4.59 \times 10^{14}} = 6.53 \times 10^{-7}$ m
(Radiation of this wavelength is red light.)

# LINE SPECTRA

## EXPLAINING LINE EMISSION SPECTRUM OF HYDROGEN

- Burning hydrogen gas or using a discharge tube filled with hydrogen excites electrons into higher energy levels.
- An electron only stays in a higher level for a short time before relaxing into lower levels.

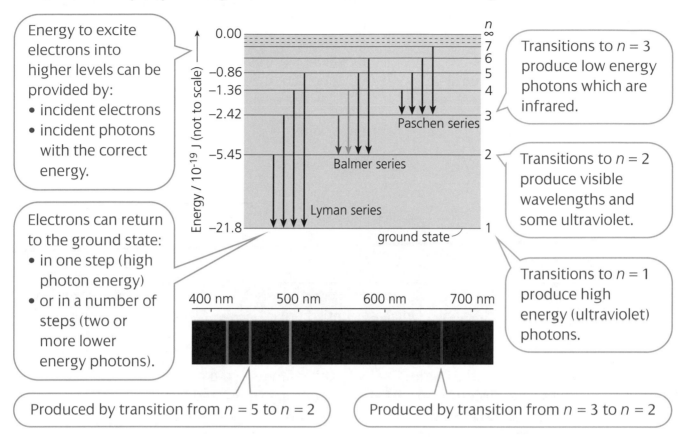

Energy to excite electrons into higher levels can be provided by:
- incident electrons
- incident photons with the correct energy.

Electrons can return to the ground state:
- in one step (high photon energy)
- or in a number of steps (two or more lower energy photons).

Transitions to $n = 3$ produce low energy photons which are infrared.

Transitions to $n = 2$ produce visible wavelengths and some ultraviolet.

Transitions to $n = 1$ produce high energy (ultraviolet) photons.

Paschen series

Balmer series

Lyman series

ground state

Produced by transition from $n = 5$ to $n = 2$

Produced by transition from $n = 3$ to $n = 2$

## EXPLAINING ABSORPTION SPECTRA

- Atoms in the cooler gas are excited by light photons that have the 'correct' energies.
- These energies are unique to the absorbing gas.

The gas atoms:
- re-emit the photons in random directions
- relax in steps emitting lower energy photons.

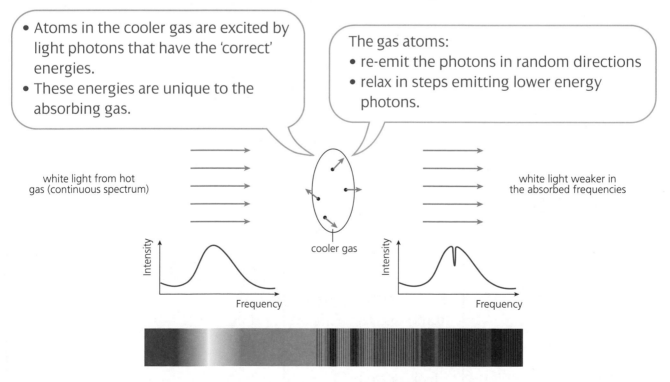

white light from hot gas (continuous spectrum)

white light weaker in the absorbed frequencies

cooler gas

Intensity

Frequency

Intensity

Frequency

- The frequencies that are absorbed by the cool gas show as dark lines in the continuous spectrum.
- The elements in the gas can be identified from these frequencies.

# LASERS

### Light Amplification by Stimulated Emission of Radiation

- For laser operation there must be a **metastable state** which leads to **population inversion**.

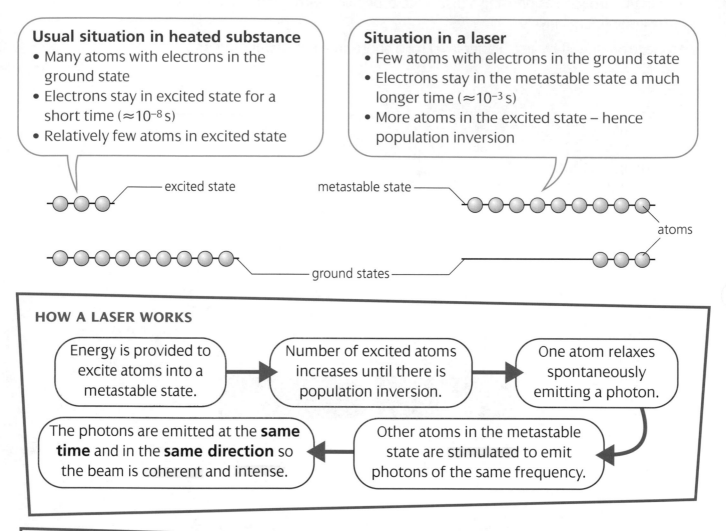

**Usual situation in heated substance**
- Many atoms with electrons in the ground state
- Electrons stay in excited state for a short time ($\approx 10^{-8}$ s)
- Relatively few atoms in excited state

**Situation in a laser**
- Few atoms with electrons in the ground state
- Electrons stay in the metastable state a much longer time ($\approx 10^{-3}$ s)
- More atoms in the excited state – hence population inversion

excited state

metastable state

atoms

ground states

## HOW A LASER WORKS

Energy is provided to excite atoms into a metastable state.

→ Number of excited atoms increases until there is population inversion.

→ One atom relaxes spontaneously emitting a photon.

Other atoms in the metastable state are stimulated to emit photons of the same frequency.

The photons are emitted at the **same time** and in the **same direction** so the beam is coherent and intense.

## OPERATION OF HELIUM-NEON LASER

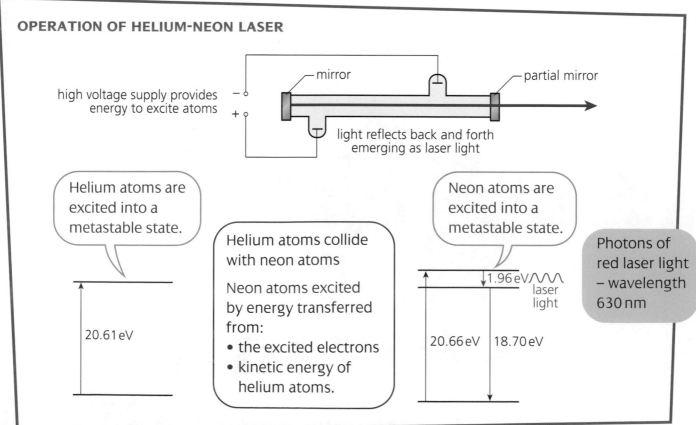

high voltage supply provides energy to excite atoms

mirror

partial mirror

light reflects back and forth emerging as laser light

Helium atoms are excited into a metastable state.

Neon atoms are excited into a metastable state.

Helium atoms collide with neon atoms

Neon atoms excited by energy transferred from:
- the excited electrons
- kinetic energy of helium atoms.

20.61 eV

1.96 eV
laser light

20.66 eV

18.70 eV

Photons of red laser light – wavelength 630 nm

# INTERNAL ENERGY AND TEMPERATURE

Solid modelled as atoms with bonds that behave like springs

- Supplying energy to matter increases internal energy of the matter.
- Internal energy is the sum of the kinetic and potential energies of all the atoms or molecules that make up the body.

- In a solid the particles can gain energy by:
  - vibrating with greater amplitude (gaining vibrational K.E.)
  - moving further apart (gaining P.E.).
- Increased internal energy is associated with a higher temperature.

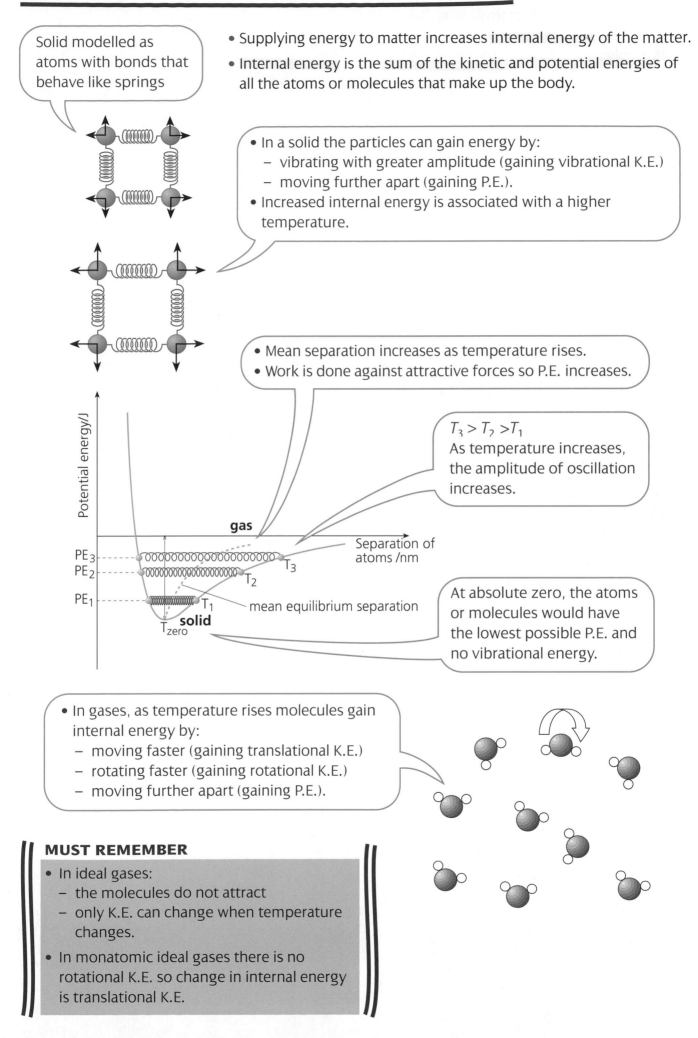

- Mean separation increases as temperature rises.
- Work is done against attractive forces so P.E. increases.

$T_3 > T_2 > T_1$
As temperature increases, the amplitude of oscillation increases.

At absolute zero, the atoms or molecules would have the lowest possible P.E. and no vibrational energy.

- In gases, as temperature rises molecules gain internal energy by:
  - moving faster (gaining translational K.E.)
  - rotating faster (gaining rotational K.E.)
  - moving further apart (gaining P.E.).

**MUST REMEMBER**

- In ideal gases:
  - the molecules do not attract
  - only K.E. can change when temperature changes.
- In monatomic ideal gases there is no rotational K.E. so change in internal energy is translational K.E.

# ZEROTH LAW OF THERMODYNAMICS

- When two bodies are in thermal equilibrium:
  - there is no net energy transfer between the two bodies
  - the bodies are at the same temperature.

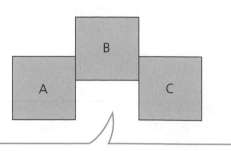

> When bodies are in contact, particles collide and exchange energy but overall there is no change in the mean or total energy of the particles in either body.

**ZEROTH LAW**

- **If** body **A** is in thermal equilibrium with **B** **and** body **B** is in thermal equilibrium with **C** **then** body **A** is in thermal equilibrium with **C**.
- Bodies in thermal equilibrium are at the same temperature.

> A **temperature difference** between two bodies in contact results in a **net energy flow from the hotter body** to the colder body until thermal equilibrium is reached. Temperature measures the 'degree of hotness'.

**MUST KNOW**

- The mean energy of atoms in a hotter body is higher than that in a cooler body. This defines the direction of energy flow.
- The total internal energy depends on the number of atoms so the cooler body may have a higher total internal energy.

# TEMPERATURE SCALES

| DEFINING TEMPERATURE ON KELVIN SCALE (K) | DEFINING TEMPERATURE ON CELSIUS SCALE (°C) |
|---|---|
| • **273.16 K** is the temperature of the **triple point** of water. | • The temperature at which ice melts is 0 °C. |
| • The **triple point** of water is the temperature at which ice, water and water vapour exist in equilibrium. | • On this scale, the triple point of water is 0.01 °C but is usually assumed to be 0 °C. |

- **Absolute zero** of temperature is:
  - the lowest possible temperature (0 K or −273.16 °C)
  - the temperature at which atoms have minimum internal energy.

**MUST TAKE CARE**

- It is incorrect to write the unit as °K.
- A **temperature difference in** K is the **same** as the **temperature difference in °C**.

**MUST REMEMBER**

Temperature in K $\approx$ Temperature in °C + 273

# CHANGING INTERNAL ENERGY

- Internal energy can be changed by heating or working.
- The change may lead to rise in temperature or change of phase (change of state).

- At different temperatures matter exists most commonly in three phases: solid, liquid and gas.
- Plasma exists at very high temperatures (such as in stars and discharge tubes) – matter then consists of ions and elementary particles (e.g. electrons and protons).

- Substance is at its **melting point**.
- Substance changes state from solid to liquid.
- Temperature remains constant.
- Energy supplied is called the **latent heat of fusion (or liquefaction)**.
- Energy provided overcomes intermolecular forces – molecules move more freely but they stay close together.

- Substance is at its **boiling point**.
- Substance changes state from liquid to gas.
- Temperature remains constant.
- Energy supplied is called the **latent heat of vaporisation**.
- Energy liberates molecules allowing them to move away from each other and to expand against the external pressure.

**Graph obtained when energy is supplied to a substance at a constant rate**

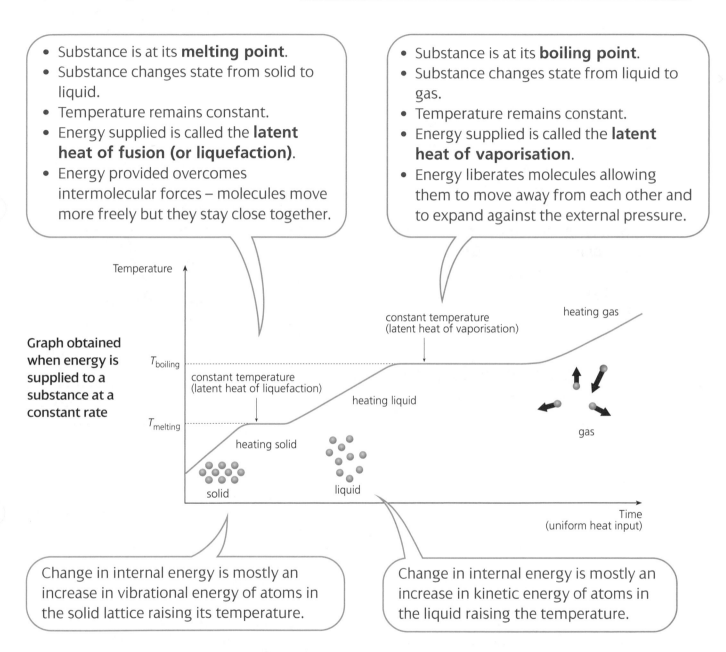

Change in internal energy is mostly an increase in vibrational energy of atoms in the solid lattice raising its temperature.

Change in internal energy is mostly an increase in kinetic energy of atoms in the liquid raising the temperature.

**MUST KNOW**

- **Specific heat capacity**, $c$, is the energy needed to raise the temperature of 1 kg of a substance by 1 K.
- **Unit of $c$ is J kg$^{-1}$ K$^{-1}$**
- **Specific latent heat of fusion**, $l_f$, is the energy needed to change 1 kg of solid into liquid when it has reached its melting point, without changing its temperature.
- **Specific latent heat of vaporisation**, $l_v$, is the energy needed to change 1 kg of liquid to gas when it has reached its boiling point, without changing its temperature.
- **Unit of $l_f$ and $l_v$ is J kg$^{-1}$**

# MEASURING SPECIFIC HEAT CAPACITY

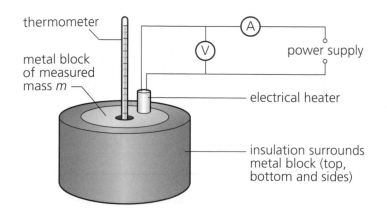

- Energy needed to raise the temperature of a mass $m$ by $\Delta\theta$, $\boldsymbol{Q = mc\Delta\theta}$
- Energy supplied in time $t$ by the electrical heater, $\boldsymbol{Q = VIt}$
- So $\boldsymbol{mc\Delta\theta = VIt}$
- Assumes all energy supplied used to heat block – some heats lagging and thermometer, and some is lost to surroundings.

- Measure mass of block.
- Measure initial temperature of the block, $\theta_i$
- Switch on supply and start timer simultaneously.
- Record readings of ammeter, $I$ and voltmeter, $V$.
- When temperature has risen about 10 degrees, stop timer and switch off simultaneously.
- Record time $t$.
- Monitor reading on thermometer and record maximum temperature reached, $\theta_f$
- Specific heat capacity is approximately,

$$c = \frac{VIt}{m(\theta_f - \theta_i)}$$

> Energy also:
> - raises temperature of heater and thermometer
> - is lost to surrounding lagging and air.
>
> This leads to too high a value for the specific heat capacity.

> For a liquid:
> - place in insulated metal calorimeter of known mass and specific heat capacity
> - measure the mass of liquid
> - stir well to distribute energy
> - remember energy $VIt$ raises temperature of liquid and calorimeter (see worked example).

**MUST TAKE CARE**

- Temperature difference in °C is equal to temperature difference in K.

  **Must not** add 273.

---

**WORKED EXAMPLE**

A copper kettle of mass 0.80 kg contains 1.20 kg of water and is heated by a 1.5 kW electric heater.
(a) Assuming no energy is lost to the surroundings and an even distribution of the energy, calculate the rise in temperature during the first 20 s of heating.
(b) How long would it take for half the water to vaporise after it has reached its boiling point?
  $c$ for copper = 390 J kg$^{-1}$ K$^{-1}$
  $c$ for water = 4200 J kg$^{-1}$ K$^{-1}$
  $l$ for water = $2.3 \times 10^6$ J kg$^{-1}$

> Most of energy supplied (94%) heats the water – copper has a relatively low specific heat capacity.

(a) Energy supplied in 20 s = 1500 × 20 = 30 000 J
  Temperature rise = $\Delta\theta$
  Energy supplied to copper in 20 s = 0.80 × 390 × $\Delta\theta$ = 312 $\Delta\theta$
  Energy supplied to water in 20 s = 1.2 × 4200 × $\Delta\theta$ = 5040 $\Delta\theta$

  312 $\Delta\theta$ + 5040 $\Delta\theta$ = 30 000 ⇒ 5352 $\Delta\theta$ = 30 000
  $\Delta\theta$ = 5.6 K

(b) Energy needed to vaporise 0.6 kg of water = $0.6 \times 2.3 \times 10^6$ J
                                              = 1 380 000 J

  At 1500 J s$^{-1}$ this takes $\dfrac{1\ 380\ 000}{1500}$ = 920 s (15.3 minutes)

# MOLAR HEAT CAPACITIES AND MOLAR LATENT HEATS

- The **molar heat capacity $C$** is the energy needed to raise the temperature of 1 mol of a substance by 1 K.

> 1 mol = $6.0 \times 10^{23}$ of atoms
> = number of atoms in the molar mass
> (e.g. 12 g for carbon, 4 g for helium)

- The **molar latent heat of fusion $L_f$** is the energy needed to melt 1 mol of a substance at its melting point without change of temperature.

- Molar heat capacity = $\dfrac{M}{1000} \times$ specific heat capacity

> Molar heat capacities and latent heats are most commonly used when working with gases.

  Molar latent heat = $\dfrac{M}{1000} \times$ specific latent heat,

  where $M$ is the molar mass in g.

# CONTINUOUS FLOW SYSTEMS

- In many applications a fluid (liquid or gas) is heated or cooled continuously while flowing through a heating or cooling chamber.

**EXAMPLES OF CONTINUOUS FLOW HEATING**
- Water heated in boiler of central heating system
- Gas heated while passing through nuclear reactor
- Water heated in electric shower

**EXAMPLES OF CONTINUOUS FLOW COOLING**
- Water cooled while passing through car radiator
- Gas cooled while passing through heat exchanger

**Example of electric shower**

> Volume $V$ flows in time $t$:
> mass = volume × density
> mass flow rate = $\dfrac{V\rho}{t}$

> energy input per second
> = power of heater $P$

mains supply

**Cold water in**
Mass of water heated per second = $\dfrac{V\rho}{t}$

Temperature of cold water = $\theta_c$

electrical heater in a water heating chamber

**Hot water out**
Temperature of hot water = $\theta_h$

Energy supplied per second $= P$

Energy gained by water per second $= \dfrac{V\rho}{t} \times c \times (\theta_h - \theta_c)$

$$P = \dfrac{V\rho}{t} \times c \times (\theta_h - \theta_c)$$

**MUST TAKE CARE**
Read question carefully.
Does it ask for temperature change or actual temperatures?

> Shows that temperature rise of water is greater for higher power heater and/or lower volume flow rate.

# MAKING ICE

- When ice is made in a refrigerator, energy is extracted from the water:
  - to reduce the temperature of water to its freezing point
  - to change the water into ice
  - to reduce the temperature of the ice to a temperature below the freezing point.

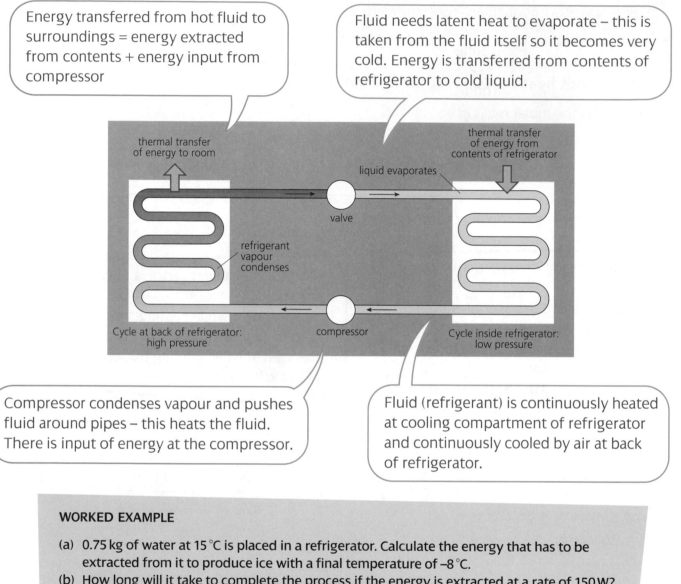

Energy transferred from hot fluid to surroundings = energy extracted from contents + energy input from compressor

Fluid needs latent heat to evaporate – this is taken from the fluid itself so it becomes very cold. Energy is transferred from contents of refrigerator to cold liquid.

thermal transfer of energy to room

thermal transfer of energy from contents of refrigerator

liquid evaporates

valve

refrigerant vapour condenses

Cycle at back of refrigerator: high pressure

compressor

Cycle inside refrigerator: low pressure

Compressor condenses vapour and pushes fluid around pipes – this heats the fluid. There is input of energy at the compressor.

Fluid (refrigerant) is continuously heated at cooling compartment of refrigerator and continuously cooled by air at back of refrigerator.

## WORKED EXAMPLE

(a) 0.75 kg of water at 15 °C is placed in a refrigerator. Calculate the energy that has to be extracted from it to produce ice with a final temperature of –8 °C.

(b) How long will it take to complete the process if the energy is extracted at a rate of 150 W?
Specific heat capacity of water = 4.2 kJ kg$^{-1}$ K$^{-1}$
Latent heat of fusion of ice = 330 kJ kg$^{-1}$
Specific heat capacity of ice = 2.1 kJ kg$^{-1}$ K$^{-1}$

(a) Energy lost when water temperature falls to freezing point $= (mc\Delta\theta)_{water}$
$$= 0.75 \times 4200 \times 15 \text{ J}$$
$$= 47\,250 \text{ J}$$

Energy lost to freeze water at its freezing point $= ml_f = 0.75 \times 330\,000$
$$= 247\,500 \text{ J}$$

Energy lost when ice temperature falls to –8 °C $= (mc\Delta\theta)_{ice} = 0.75 \times 2100 \times 8 \text{ J}$
$$= 12\,600 \text{ J}$$

Total energy extracted $= 47\,250 + 247\,500 + 12\,600 = 307 \text{ kJ}$

(b) Time taken $= \dfrac{\text{energy extracted}}{\text{rate of extracting energy}} = \dfrac{307\,000}{150}$

$$= 2050 \text{ s} = 34.1 \text{ minutes}$$

# GAS LAWS

- The state of a gas is defined by its pressure $p$, volume $V$ and temperature $T$.
- The gas laws summarise the behaviour of an **ideal gas**.
- An ideal gas is one that obeys **Boyle's law** at all temperatures.

## BOYLE'S LAW

**For a fixed mass of gas at constant temperature the pressure is inversely proportional to the volume:**

$$p \propto \frac{1}{V}$$

$$pV = \text{constant}$$

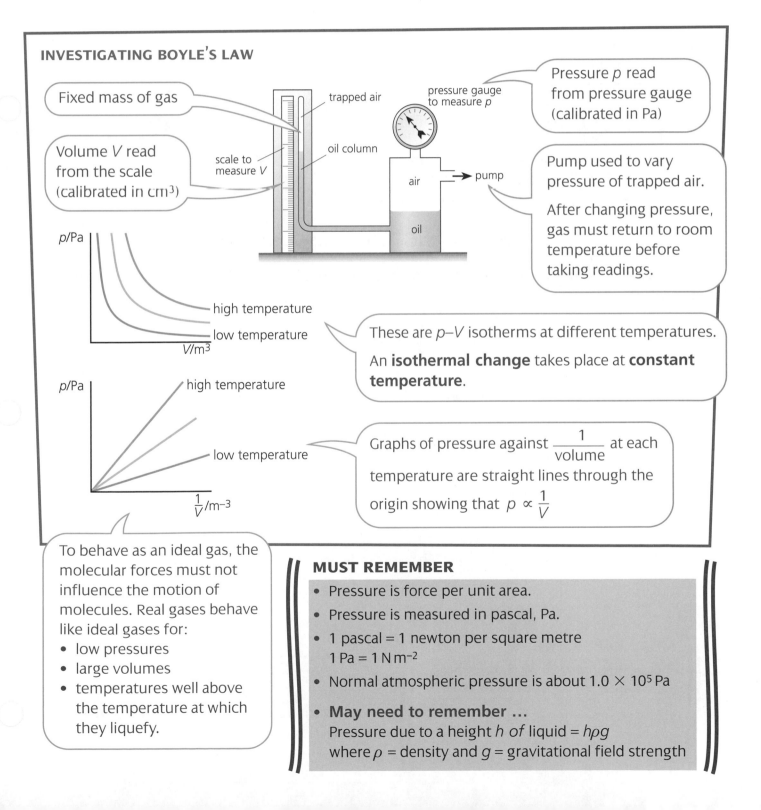

**INVESTIGATING BOYLE'S LAW**

Fixed mass of gas

Volume $V$ read from the scale (calibrated in cm³)

scale to measure $V$

trapped air

oil column

air

oil

pressure gauge to measure $p$

pump

Pressure $p$ read from pressure gauge (calibrated in Pa)

Pump used to vary pressure of trapped air.

After changing pressure, gas must return to room temperature before taking readings.

$p$/Pa

high temperature

low temperature

$V$/m³

These are $p$–$V$ isotherms at different temperatures.

An **isothermal change** takes place at **constant temperature**.

$p$/Pa

high temperature

low temperature

$\frac{1}{V}$/m⁻³

Graphs of pressure against $\frac{1}{\text{volume}}$ at each temperature are straight lines through the origin showing that $p \propto \frac{1}{V}$

To behave as an ideal gas, the molecular forces must not influence the motion of molecules. Real gases behave like ideal gases for:
- low pressures
- large volumes
- temperatures well above the temperature at which they liquefy.

**MUST REMEMBER**
- Pressure is force per unit area.
- Pressure is measured in pascal, Pa.
- 1 pascal = 1 newton per square metre
  1 Pa = 1 N m⁻²
- Normal atmospheric pressure is about $1.0 \times 10^5$ Pa
- **May need to remember …**
  Pressure due to a height $h$ of liquid = $h\rho g$
  where $\rho$ = density and $g$ = gravitational field strength

# INVESTIGATING CHARLES' LAW

**For a fixed mass of gas at constant pressure the volume is proportional to the absolute temperature.**

$$V \propto T$$
$$\frac{V}{T} = constant$$

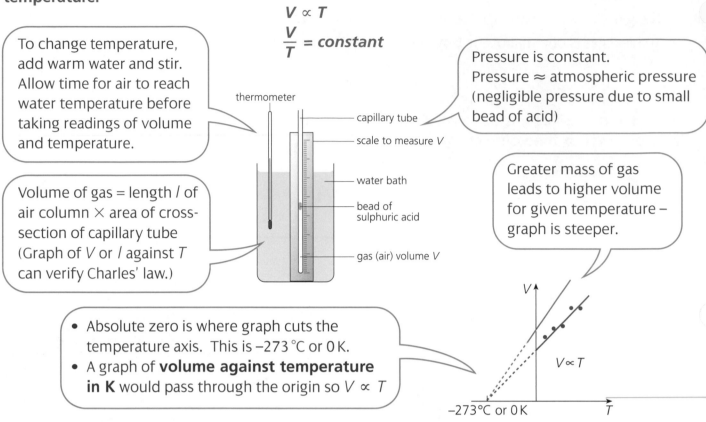

To change temperature, add warm water and stir. Allow time for air to reach water temperature before taking readings of volume and temperature.

Pressure is constant. Pressure ≈ atmospheric pressure (negligible pressure due to small bead of acid)

Volume of gas = length *l* of air column × area of cross-section of capillary tube (Graph of *V* or *l* against *T* can verify Charles' law.)

Greater mass of gas leads to higher volume for given temperature – graph is steeper.

thermometer
capillary tube
scale to measure *V*
water bath
bead of sulphuric acid
gas (air) volume *V*

- Absolute zero is where graph cuts the temperature axis. This is −273 °C or 0 K.
- A graph of **volume against temperature in K** would pass through the origin so $V \propto T$

$V \propto T$

−273°C or 0 K

# INVESTIGATING THE PRESSURE LAW

**The pressure of a fixed mass of gas at constant volume is proportional to the absolute temperature.**

$$p \propto T$$
$$\frac{p}{T} = constant$$

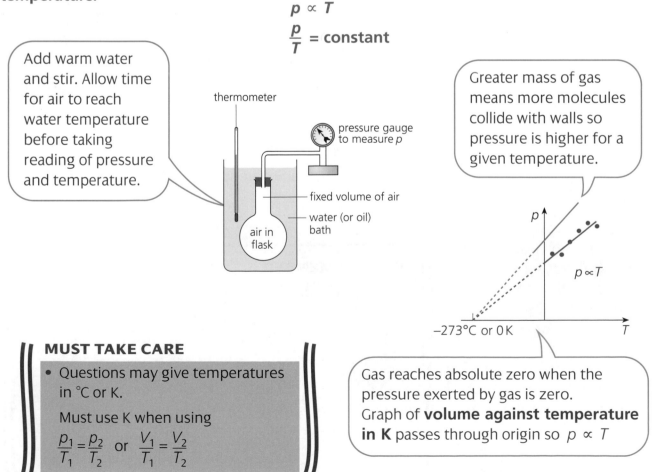

Add warm water and stir. Allow time for air to reach water temperature before taking reading of pressure and temperature.

Greater mass of gas means more molecules collide with walls so pressure is higher for a given temperature.

thermometer
pressure gauge to measure *p*
fixed volume of air
water (or oil) bath
air in flask

$p \propto T$

−273°C or 0 K

**MUST TAKE CARE**

- Questions may give temperatures in °C or K.

  Must use K when using

  $$\frac{p_1}{T_1} = \frac{p_2}{T_2} \quad \text{or} \quad \frac{V_1}{T_1} = \frac{V_2}{T_2}$$

Gas reaches absolute zero when the pressure exerted by gas is zero. Graph of **volume against temperature in K** passes through origin so $p \propto T$

# IDEAL GAS EQUATIONS

Combining Boyle's Law, Charles' law and the pressure law gives the **ideal gas equation**.

$$\frac{pV}{T} = \text{constant}$$

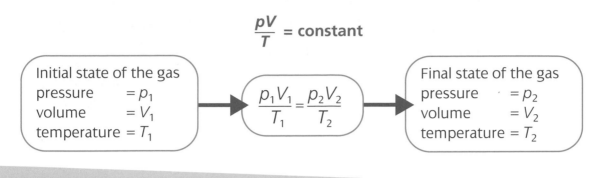

Initial state of the gas
pressure $= p_1$
volume $= V_1$
temperature $= T_1$

$$\frac{p_1 V_1}{T_1} = \frac{p_2 V_2}{T_2}$$

Final state of the gas
pressure $= p_2$
volume $= V_2$
temperature $= T_2$

## WORKED EXAMPLE

Air in a container has a volume of 500 cm³. Initially its pressure is $0.98 \times 10^5$ Pa and temperature is 27 °C. It is expanded to 700 cm³ and the pressure decreases to $0.65 \times 10^5$ Pa. Calculate the final temperature, in °C, of the air in the container.

> Avoid mistakes – list initial and final conditions of gas.

$p_1 = 0.98 \times 10^5$ Pa        $p_2 = 0.65 \times 10^5$ Pa
$V_1 = 500$ cm³                 $V_2 = 700$ cm³
$T_1 = 300$ K                   $T_2 = ?$

$$\frac{p_1 V_1}{T_1} = \frac{p_2 V_2}{T_2} \Rightarrow \frac{0.98 \times 10^5 \times 500}{300} = \frac{0.65 \times 10^5 \times 700}{T_2}$$

$$T_2 = \frac{1.365 \times 10^{10}}{4.9 \times 10^7} = 278.6 \text{ K}$$

Final temperature $= 278.6 - 273 = 5.6$ °C

**MUST REMEMBER**

When using the ideal gas equation:

- units of pressure and volume must be the same on both sides of the equation

- **Temperatures must be in kelvin, K.**

# UNIVERSAL GAS EQUATION

- Pressure of a gas is:
  - proportional to $T$
  - inversely proportional to $V$
  - proportional to the amount of gas present, $n$

> **Amount** of gas is measured in **moles**. 1 mol of a substance contains $6.02 \times 10^{23}$ atoms or molecules.

$$p \propto \frac{nT}{V} \quad \text{so} \quad pV = \text{constant} \times nT$$

$$pV = nRT$$

- $R$ is the **universal molar gas constant**
- $R = 8.3$ J K⁻¹ mol⁻¹

**MUST TAKE CARE**

- **$n$ is number of moles not** number of atoms or molecules present

- $p$ must be in Pa

- $V$ must be in m³

## WORKED EXAMPLE

Helium gas in a 1 litre container has pressure of $1.0 \times 10^5$ Pa and temperature of 300 K.
Calculate:
(a) the number of atoms present in a litre of helium gas at room temperature
(b) the volume that contains one mol (4 g) of helium at this temperature and pressure.

(a) $pV = nRT$
1 litre $= 1000$ cm³ $= 1000 \times 10^{-6}$ m³
$1.0 \times 10^5 \times 1000 \times 10^{-6} = n \times 8.3 \times 300$
Amount of helium present $n = 0.0402$ mol
Number of atoms present $= 0.0402 \times 6.02 \times 10^{23}$
$= 2.4 \times 10^{22}$

(b) $1.0 \times 10^5 \times V = 1 \times 8.3 \times 300$
$V = 0.0249$ m³ (24.9 litres)

# THERMOMETERS

## IDEAL GAS TEMPERATURE SCALE

- From $pV = nRT$ for a fixed amount of gas, $pV \propto T$
- Two fixed points are needed to define the temperature scale.

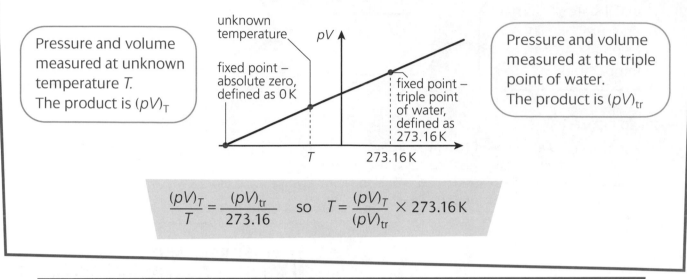

Pressure and volume measured at unknown temperature $T$. The product is $(pV)_T$

unknown temperature

fixed point – absolute zero, defined as 0 K

fixed point – triple point of water, defined as 273.16 K

Pressure and volume measured at the triple point of water. The product is $(pV)_{tr}$

$$\frac{(pV)_T}{T} = \frac{(pV)_{tr}}{273.16} \quad \text{so} \quad T = \frac{(pV)_T}{(pV)_{tr}} \times 273.16\,\text{K}$$

## CONSTANT VOLUME GAS THERMOMETER

- If $V$ is kept constant then the ideal gas temperature scale becomes

$$T = \frac{p_T}{p_{tr}} \times 273.16\ \text{K}$$

- The constant volume gas thermometer is the standard against which other thermometers are calibrated but it has too many disadvantages for normal use.

## COMPARISON OF DIFFERENT TYPES OF THERMOMETER

| Thermometer | Thermometric property | Main advantages | Main disadvantages | Temperature range |
|---|---|---|---|---|
| Liquid-in-glass | Volume change (i.e. changing length of thread of mercury or alcohol) | Simple to use, cheap, portable | Fragile, limited range, not suitable for small objects | Mercury: 234–723 K; Ethanol: 173–323 K |
| Constant-volume gas thermometer | Pressure of fixed mass of gas at constant volume | Absolute scale given, accurate, wide range | Bulky and inconvenient, slow response, not suitable for small objects directly | 3–500 K |
| Resistance | Resistance of platinum wire | Accurate, wide range, useful for small temperature differences | Slow to respond, not suitable for small objects | 15–900 K |
| Thermocouple | E.m.f. across junction of two dissimilar metals | Can measure small differences, fast response, wide range, can be read remotely | Small voltages, so need electronic amplification | 25–1400 K (depending on metals) |
| Thermistor | Changing resistance of semiconductor | Provides electrical signal suitable for computer circuits | Calibration necessary, not very accurate | 200–700 K |
| Optical pyrometer | Adjustment of current through lamp filament to match colour of object | No contact with hot object, simple to use, portable | Calibration necessary, not very accurate | Above 1250 K |

# KINETIC THEORY OF GASES

- **Kinetic theory** assumes the existence of fast-moving molecules moving randomly.
- **Brownian motion** provides evidence for this assumption.

**Brownian motion** is the random motion of smoke particles, viewed under a microscope. (In original experiment pollen was suspended in a liquid.)

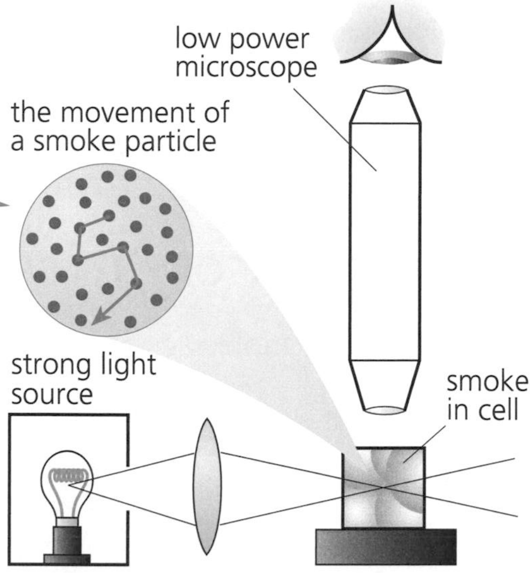

low power microscope

the movement of a smoke particle

strong light source

smoke in cell

**EXPLAINING BROWNIAN MOTION**

- Smoke particles are bombarded by air molecules.
- Air molecules have momentum and exert a force on the smoke particle.
- The collisions are random and the smoke particles are small.
- At any instant more collisions produce a force in one direction than in the opposite direction.
- The direction of the unbalanced force continually changes direction.
- The unbalanced forces produce the jerky random motion.

- Smoke is in a smoke cell to eliminate the effects of draughts.
- It is viewed using light reflected from the particles.

## WHAT ARE THE ASSUMPTIONS OF THE KINETIC THEORY?

- A gas consists of a very large number of molecules.
- Molecules move rapidly and randomly.
- Collisions between the molecules and the walls of the container are perfectly elastic.
- Molecules do not attract each other.
- Only forces between molecules are those that occur during collisions.
- Duration of a collision is small compared with the time between collisions.
- Volume of a molecule is negligible compared with that occupied by the gas.

**MUST KNOW**

Popular in exam questions – must know the assumptions

**KINETIC THEORY SUCCESSES**

- Kinetic theory can explain:
  - gas pressure
  - variation of pressure with volume and temperature
  - diffusion of gases.

# EXPLAINING GAS PRESSURE

- Molecules in constant rapid motion have **momentum**.
- When they collide with the walls of the vessel the momentum changes.
- The wall exerts a force on a molecule to change its momentum.
- The wall experiences an equal and opposite force (Newton's 3rd Law).
- Lots of molecules collide with the wall so there is a large force on the wall.
- The force is equal to the momentum change each second.
- Pressure is force exerted per square metre, $p = \dfrac{F}{A}$

## USING KINETIC THEORY TO SHOW THAT $pV = \frac{1}{3}Nm\langle c^2 \rangle$ AND $p = \frac{1}{3}\rho\langle c^2 \rangle$

- Suppose there are $N$ molecules of mass $m$ in a cubic box of side $l$.
- Taking a simplified view, there are $\dfrac{N}{3}$ molecules moving with a speed $u$ in each of the directions $x$, $y$ and $z$. ($u = \sqrt{\langle c^2 \rangle}$, the root mean square speed of the molecules)

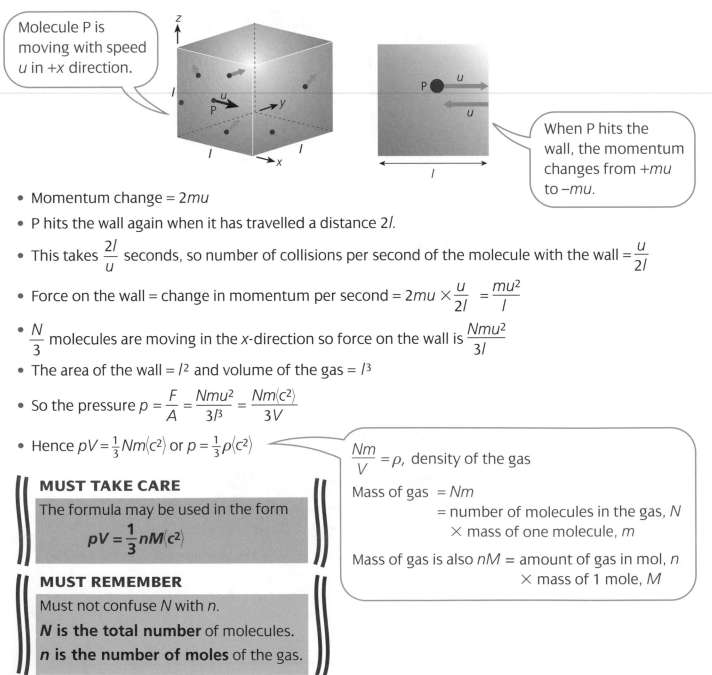

Molecule P is moving with speed $u$ in $+x$ direction.

When P hits the wall, the momentum changes from $+mu$ to $-mu$.

- Momentum change $= 2mu$
- P hits the wall again when it has travelled a distance $2l$.
- This takes $\dfrac{2l}{u}$ seconds, so number of collisions per second of the molecule with the wall $= \dfrac{u}{2l}$
- Force on the wall $=$ change in momentum per second $= 2mu \times \dfrac{u}{2l} = \dfrac{mu^2}{l}$
- $\dfrac{N}{3}$ molecules are moving in the $x$-direction so force on the wall is $\dfrac{Nmu^2}{3l}$
- The area of the wall $= l^2$ and volume of the gas $= l^3$
- So the pressure $p = \dfrac{F}{A} = \dfrac{Nmu^2}{3l^3} = \dfrac{Nm\langle c^2 \rangle}{3V}$
- Hence $pV = \frac{1}{3}Nm\langle c^2 \rangle$ or $p = \frac{1}{3}\rho\langle c^2 \rangle$

$\dfrac{Nm}{V} = \rho$, density of the gas

Mass of gas $= Nm$
$\qquad$ = number of molecules in the gas, $N$
$\qquad \times$ mass of one molecule, $m$

Mass of gas is also $nM$ = amount of gas in mol, $n$
$\qquad \times$ mass of 1 mole, $M$

**MUST TAKE CARE**

The formula may be used in the form

$$pV = \frac{1}{3}nM\langle c^2 \rangle$$

**MUST REMEMBER**

Must not confuse $N$ with $n$.

**$N$ is the total number** of molecules.

**$n$ is the number of moles** of the gas.

# USING KINETIC THEORY TO EXPLAIN GAS LAWS

## EXPLAINING BOYLE'S LAW

**Why does the pressure of a gas increase when volume decreases?**

- If volume is reduced molecules have less distance to travel between collisions with a wall of the container.
- They collide with the walls more often.
- There is a greater change in momentum per second.
- There is a greater force and a greater pressure.

**MUST KNOW**

Must know these explanations – they are often asked for in exams

## EXPLAINING CHARLES' LAW

**Why does the volume increase with temperature when pressure is constant?**

- Increased temperature increases the molecular speed.
  - The momentum of each molecule increases.
  - Force for each collision increases.
  - To keep pressure constant there must be fewer collisions with the walls each second.
  - The molecules need to move further between collisions.
  - This is achieved by increasing volume.

## EXPLAINING THE PRESSURE LAW

**Why does the pressure increase with temperature when volume is constant?**

- Increased temperature increases molecular speeds.
  - The momentum of each molecule increases.
  - Each molecule collides more frequently with the walls of the container.
  - Each of these effects increases the momentum change per second.
  - There is a greater force and therefore a greater pressure.

# HOW FAST DO MOLECULES MOVE?

- Vertical axis represents number of molecules with similar speeds.
- At high temperatures there are more molecules with high speeds and mean speed is higher.
- Even at low temperatures some molecules have a high speed.

- Molecules of a gas have a range of speeds (and therefore a range of kinetic energies).
- The distribution of the speeds varies with temperature.

## ROOT MEAN SQUARE SPEED

- When applying the kinetic theory (see previous page) the **root mean square speed** is important.

  This is $\sqrt{\langle c^2 \rangle} = \sqrt{\dfrac{c_1^2 + c_2^2 + \ldots + c_n^2}{n}}$

  where $c_1$, $c_2$, etc. are speeds of individual molecules in gas and $n$ = number of molecules

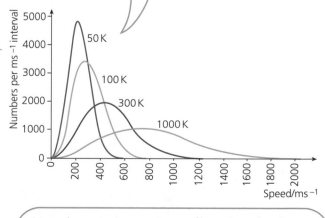

At a given temperature all molecules have the same mean K.E. so lighter molecules have higher mean speeds.

# INTERNAL ENERGY OF AN IDEAL GAS

- The **internal energy $U$** of an ideal gas is the kinetic energy of the atoms and molecules in it.

  - Kinetic energy of a molecule $= \frac{1}{2}m\langle c^2\rangle$

  - Total internal energy of $N$ molecules $= \frac{1}{2}Nm\langle c^2\rangle$

  - From kinetic theory, $pV = \frac{1}{3}Nm\langle c^2\rangle$

  - So $Nm\langle c^2\rangle = 3pV$ and $\frac{1}{2}Nm\langle c^2\rangle = \frac{3}{2}pV$

  - From gas laws, $pV = nRT$ where $n$ is the amount of gas in mol

- Total internal energy of $N$ molecules, $U = \frac{3}{2}nRT$

> **MUST REMEMBER**
>
> There is no internal potential energy as the molecules of an ideal gas do not attract each other.

> See page 44 for origins of these equations.

> Amount of gas, $n = \dfrac{N}{N_A}$ mol
>
> Avogadro constant $N_A$ = number of molecules in 1 mol of a substance
>
> so $N$ atoms $\equiv \dfrac{N}{N_A}$ mol of gas

**WORKED EXAMPLE**

0.12 g of helium is contained in a vessel of volume 750 cm³.
Calculate:
(a)  the pressure exerted by the helium when the atoms have a mean speed of 1500 m s⁻¹
(b)  the total internal energy of the helium atoms
(c)  the temperature of the gas.

Molar mass of helium = 4 g
Gas constant = 8.3 J K⁻¹ mol⁻¹

(a)  $pV = \frac{1}{3}Nm\langle c^2\rangle \Rightarrow p = \frac{1}{3}\dfrac{Nm}{V}\langle c^2\rangle$

$Nm$ = total mass of helium

$p = \frac{1}{3} \times \dfrac{0.000\,12}{750 \times 10^{-6}} \times 1500^2 = 1.2 \times 10^5\,\text{Pa}$

(b)  Total internal energy = Number of atoms × mean K.E. per atom = $N \times \frac{1}{2}m\langle c^2\rangle$

$\frac{1}{2}Nm\langle c^2\rangle = \frac{1}{2} \times 0.000\,12 \times 1500^2 = 135\,\text{J}$

(c)  $pV = nRT$ where $n = \dfrac{\text{mass of gas}}{\text{molar mass}}$

$T = \dfrac{pV}{nR} = \dfrac{1.2 \times 10^5 \times 750 \times 10^{-6}}{\frac{0.12}{4} \times 8.3} = 361\,\text{K}$

> **MUST TAKE CARE**
>
> To convert cm³ to m³ multiply by 10⁻⁶

# BOLTZMANN CONSTANT $k$

> $k$ = Boltzmann constant
>
> $k = \dfrac{8.3}{6.02 \times 10^{23}} = 1.38 \times 10^{-23}\,\text{J K}^{-1}$

- **Mean kinetic energy of $N$ molecules $= \frac{3}{2}\dfrac{N}{N_A}RT$**

- **Mean kinetic energy of 1 molecule $= \frac{3}{2}\dfrac{R}{N_A}T = \frac{3}{2}kT$**

> This equation gives mean molecular K.E. at any temperature.
> Hence mean speed of molecules can be found.

> This is approximately the same speed that the peak of the graph for 300 K occurs on page 45.

**WORKED EXAMPLE**

Calculate the mean speed of oxygen molecules at 300 K.
Mass of an oxygen molecule = $5.4 \times 10^{-26}$ kg

Mean K.E. $= \frac{3}{2}kT$

$\qquad = \frac{3}{2} \times 1.38 \times 10^{-23} \times 300$

$\qquad = 6.21 \times 10^{-21}\,\text{J}$

$\frac{1}{2}mv^2 = 6.21 \times 10^{-21}$

$v = 480\,\text{m s}^{-1}$

# FIRST LAW OF THERMODYNAMICS

- There are two ways of changing internal energy of a system. It can be heated or work can be done on it.

$$\Delta U = Q + W$$

Change in internal energy = energy supplied by heating + energy supplied by doing work

surroundings

SYSTEM
Change in internal energy $\Delta U$

system boundary

Energy $Q$ supplied **to** the system by heating

Energy $W$ supplied by doing work **on** the system (by working)

Energy transferred by **heating** when system is in contact with hotter system such as Bunsen flame or electric heater.

Energy transferred by **doing work** when:
- a gas is compressed or expanded
- a solid is subjected to frictional forces or is struck (with a hammer)
- an object hits the ground having fallen under gravity.

## WHAT IS THE SYSTEM?

- A system is all the matter within a defined boundary.

- Everything outside the boundary is the surroundings.

- A system may be:
  - a solid object
  - a defined mass of liquid or gas
  - a number of different lumps of matter with a defined boundary.

- The system can:
  - gain energy by being heated by the surroundings or having work done on it by the surroundings
  - lose energy by supplying energy to the surroundings by heating it or by doing work on the surroundings.

- The system boundaries must be identified when applying the first law.

- The first law can be applied to the system or to the surroundings.

## MUST TAKE CARE

The law is sometimes stated as

$$\Delta U = Q - W$$

In this case the system **loses** internal energy by doing work **on** the surroundings.

The **universe** is the **system** being considered plus the **surroundings**, i.e. everything else.

## MUST REMEMBER

- When there is an increase in internal energy $\Delta U$, there is a rise in temperature:

$\Delta\theta = \dfrac{\Delta U}{mc}$, where $m$ is the mass and $c$ the specific heat capacity of the system.

- A rise in temperature can be produced by heating or by doing work.

# EXPANDING AND COMPRESSING GASES

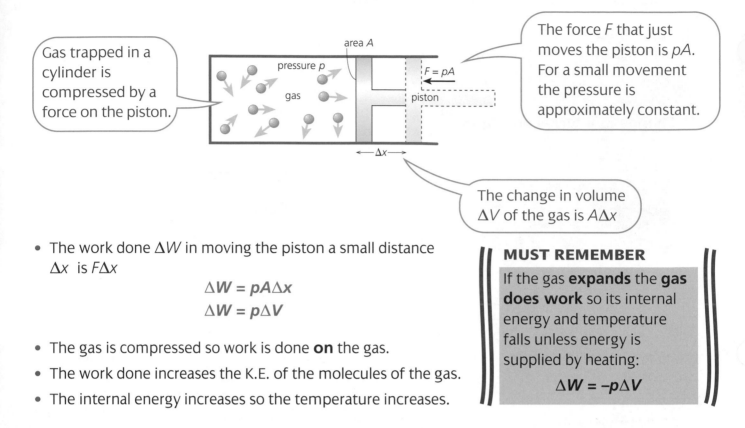

Gas trapped in a cylinder is compressed by a force on the piston.

area $A$

pressure $p$

gas

$F = pA$

piston

The force $F$ that just moves the piston is $pA$. For a small movement the pressure is approximately constant.

$\leftarrow \Delta x \rightarrow$

The change in volume $\Delta V$ of the gas is $A\Delta x$

- The work done $\Delta W$ in moving the piston a small distance $\Delta x$ is $F\Delta x$

$$\Delta W = pA\Delta x$$
$$\Delta W = p\Delta V$$

- The gas is compressed so work is done **on** the gas.
- The work done increases the K.E. of the molecules of the gas.
- The internal energy increases so the temperature increases.

**MUST REMEMBER**

If the gas **expands** the **gas does work** so its internal energy and temperature falls unless energy is supplied by heating:

$$\Delta W = -p\Delta V$$

## WORK DONE WHEN VOLUME CHANGES AT CONSTANT PRESSURE

- Since pressure is constant in this change,
  **work done = pressure × change in volume**

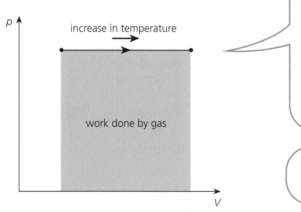

$p$

increase in temperature

work done by gas

$V$

- This is an **isobaric** change.
- Work is done **by** the gas.  (**$W$ is negative**)
- Temperature increases.   (**$V \propto T$**)
- Energy is supplied by heating to maintain original pressure.        (**$Q$ is positive**)
- Internal energy increases. (**$\Delta U$ is positive**)

Since temperature rises, more energy is supplied as heat than the work done by the gas.

- Work done is always the area between the line and the volume axis.
- If volume increases, work is done by the gas.
- If volume decreases, work is done on the gas.
- The arrow points in the direction of the change.

**MUST TAKE CARE**

- For the change at constant pressure:
  work done =  pressure × change in volume
- If work done is in joule (J):
  pressure must be in pascal (Pa)
  volume must be in metre³ (m³).

**WORKED EXAMPLE**

Gas at a pressure of 130 kPa is expanded from 300 cm³ to 450 cm³ at constant pressure. Calculate the work done.

Work done $= p\Delta V$
$= 130 \times 10^3 \times (450 - 300) \times 10^{-6}$
$= 19.5$ J

# ISOTHERMAL AND ADIABATIC CHANGES

### ISOTHERMAL CHANGE

- No change in temperature so no change in internal energy: $\Delta U = 0$
- All energy supplied by heating used to do work on surroundings: $Q = -W$
- An isothermal change takes place slowly in a container that is a good thermal conductor.

### ADIABATIC CHANGE

- No exchange of thermal energy between body and surroundings (i.e. no heating or cooling): $Q = 0$
- All energy supplied to a system by doing work becomes internal energy of the system: $\Delta U = W$
- An adiabatic change takes place quickly in a container that is a good thermal insulator.

**Isothermal change** obeys Boyle's law: **$pV = \text{constant}$**

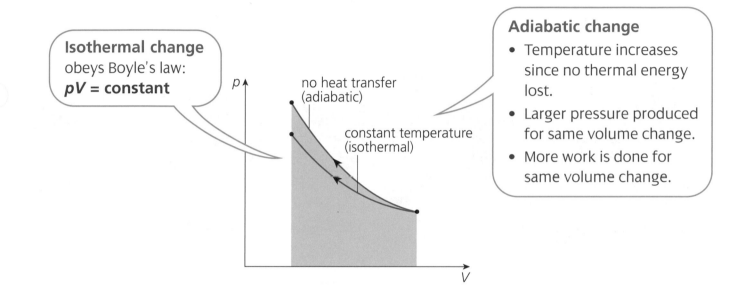

no heat transfer (adiabatic)

constant temperature (isothermal)

**Adiabatic change**
- Temperature increases since no thermal energy lost.
- Larger pressure produced for same volume change.
- More work is done for same volume change.

# CONSTANT VOLUME CHANGE

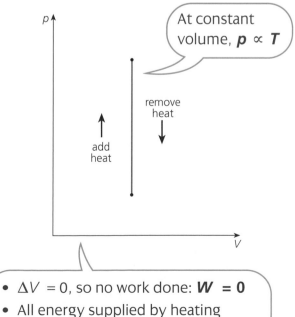

At constant volume, **$p \propto T$**

remove heat

add heat

- $\Delta V = 0$, so no work done: **$W = 0$**
- All energy supplied by heating becomes internal energy: $\Delta U = Q$

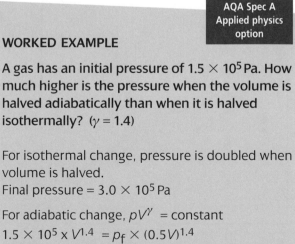

**AQA Spec A Applied physics option**

**WORKED EXAMPLE**

A gas has an initial pressure of $1.5 \times 10^5$ Pa. How much higher is the pressure when the volume is halved adiabatically than when it is halved isothermally? $(\gamma = 1.4)$

For isothermal change, pressure is doubled when volume is halved.
Final pressure = $3.0 \times 10^5$ Pa

For adiabatic change, $pV^\gamma = \text{constant}$
$1.5 \times 10^5 \times V^{1.4} = p_f \times (0.5V)^{1.4}$
$V^{1.4}$ on each side cancels so
$\qquad 1.5 \times 10^5 = p_f \times (0.5)^{1.4}$
Final pressure $p_f = \dfrac{1.5 \times 10^5}{0.379} = 3.96 \times 10^5$

Pressure is $0.96 \times 10^5$ Pa higher

# FINDING WORK DONE WHEN PRESSURE CHANGES

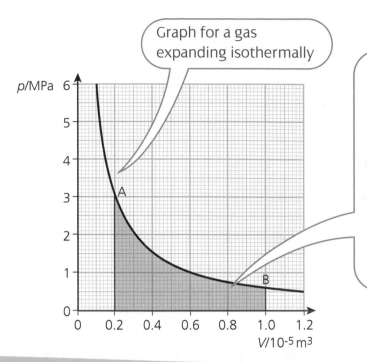

Graph for a gas expanding isothermally

- Shaded area represents work done by gas when it expands.
- To find approximate value for work done:
  - count the squares between line and V-axis. Counting smaller squares gives more reliable value. Need to judge fractions of squares.
  - calculate energy represented by one square
  - multiply number of squares by energy per square.

**WORKED EXAMPLE**

**Use the isothermal in the above diagram to estimate the work done when the gas expands from A to B.**

Number of large squares $\approx 5$
Energy represented by one large square $= (1 \times 10^6) \times (0.2 \times 10^{-5})$
$\qquad\qquad\qquad\qquad\qquad\qquad\qquad = 2\,J$
Work done $= 5 \times 2 = 10\,J$

**MUST TAKE CARE**

- The $p$–$V$ graph may be drawn with a false origin for the pressure axis.
- Need to account for this when finding area under graph

# GAS CYCLES

- A gas is taken through a cycle of changes in an engine or a refrigerator.

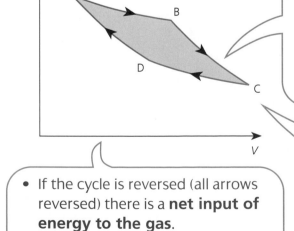

- This is a **Carnot cycle**.
- An ideal engine would use this cycle – it is the most efficient engine cycle possible.
- A→B is an **isothermal expansion** (gas does work).
  B→C is an **adiabatic expansion** (gas does work and cools).
  C→D is an **isothermal compression** (gas has work done on it).
  D→A is an **adiabatic compression** (gas does work and gets hotter).

- Area under ABC > Area under CDA
  Work done by the gas > Work done on the gas
  There is net energy output.
- **The area enclosed by the cycle is the energy output of the engine for one cycle.**

- If the cycle is reversed (all arrows reversed) there is a **net input of energy to the gas**.
- The surroundings become cooler so it works as a **refrigerator**.

For further work on gas cycles see pages 115–116

# STRESS, STRAIN AND YOUNG MODULUS

## TENSILE STRESS

- $\sigma = \dfrac{\text{tensile force}}{\text{area of cross-section}} = \dfrac{F}{A}$
- Unit is Pa or N m$^{-2}$
- Produced by equal and opposite stretching forces

## TENSILE STRAIN

- $\varepsilon = \dfrac{\text{extension}}{\text{original length}} = \dfrac{\Delta L}{L}$
- Has no unit – it is a ratio of two lengths

Force = F

Area of cross-section = A

$L$

$x$

Force = F

## YOUNG MODULUS

- $E = \dfrac{\text{stress}}{\text{strain}} = \dfrac{F/A}{\Delta L/L} = \dfrac{FL}{A\Delta L}$
- Unit is Pa or N m$^{-2}$
- Measure of stiffness
- Applies only over initial linear elastic region

Formulae for stress, strain and Young modulus can be used for **small compressions** of a rod or bar.

## MUST TAKE CARE

- Must use cross-sectional area.
- May be given area, radius or diameter.
- Area of cross section $= \pi r^2 = \dfrac{\pi d^2}{4}$
- Original length of wire $= L$
- Extension $= \Delta L$

Note: $e$ or $x$ is sometimes used instead of $\Delta L$

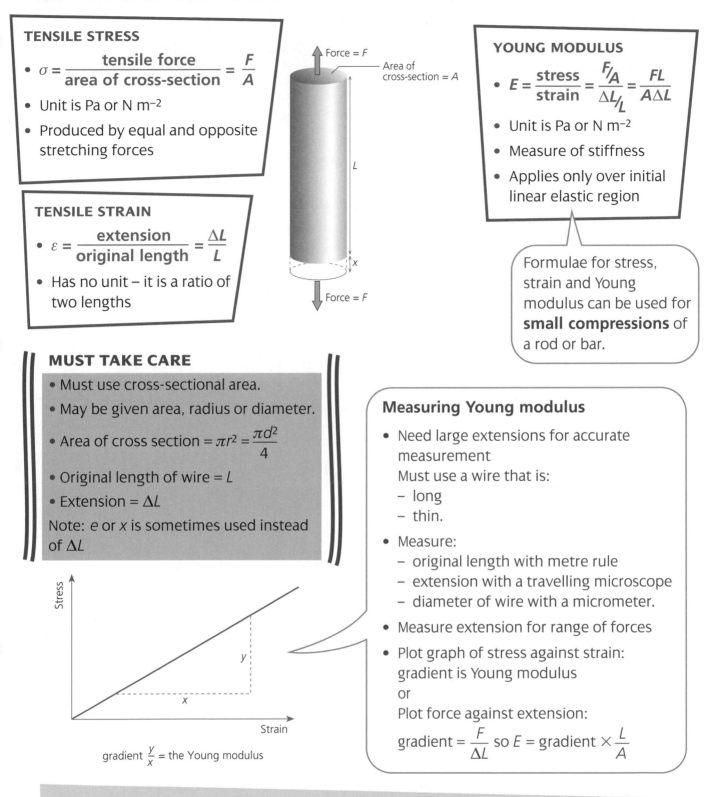

Stress

$y$

$x$

Strain

gradient $\dfrac{y}{x}$ = the Young modulus

## Measuring Young modulus

- Need large extensions for accurate measurement
  Must use a wire that is:
  – long
  – thin.
- Measure:
  – original length with metre rule
  – extension with a travelling microscope
  – diameter of wire with a micrometer.
- Measure extension for range of forces
- Plot graph of stress against strain: gradient is Young modulus
  or
  Plot force against extension:
  gradient $= \dfrac{F}{\Delta L}$ so $E$ = gradient $\times \dfrac{L}{A}$

## WORKED EXAMPLE

A wire has a length of 5.0 m and diameter 1.5 mm. The Young modulus of the wire is $1.2 \times 10^{11}$ Pa. Calculate the extension of the wire when the tensile force applied is 7.5 N.

Area of cross-section $= 3.14 \times \left(\dfrac{1.5 \times 10^{-3}}{2}\right)^2 = 1.77 \times 10^{-6}$ m$^2$

Stress $= \dfrac{F}{A} = \dfrac{7.5}{1.77 \times 10^{-6}} = 4.24 \times 10^6$ Pa

Strain $= \dfrac{\text{stress}}{E} = \dfrac{4.24 \times 10^6}{1.2 \times 10^{11}} = 3.5 \times 10^{-5}$

Extension = original length $\times$ strain $= 3.5 \times 10^{-5} \times 5.0$ m $= 1.8 \times 10^{-4}$ m

# DOING WORK ON WIRES

## ENERGY STORED IN STRETCHED SPRINGS AND WIRES

- **Work done on wire when stretching it = energy transferred to wire**
- Energy supplied can:
  - be stored as elastic potential energy
  - be used to deform the wire permanently by rearranging atoms in wire
  - become internal energy of the wire.
- Elastic P.E. is stored provided elastic limit is not exceeded.

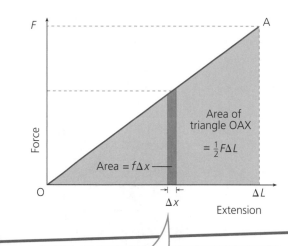

- The total work done when stretching a wire is the pale blue area under the graph line between 0 and A.

  **Work done**

  $= \frac{1}{2} \times$ **maximum force** $\times$ **maximum extension**

  $= \frac{1}{2} F \Delta L$

- Work done extending wire by $\Delta x$ when force is $f$:
  - $\approx f \Delta x$
  - $\approx$ area of dark blue strip
- Total work done = sum of all similar strips that make up pale blue area.

## MUST REMEMBER

- Work done is in J if:
  - $F$ is in N
  - $\Delta L$ is in m
- If extension produced by suspending mass $m$ kg:
  $$\text{final force} = mg$$
  $$\text{tension in wire} = mg$$
- For a curved graph, find work done by counting squares and multiplying by work done per square.

## WORKED EXAMPLE

A steel wire of diameter 2.0 mm and length 2.5 m extends by 0.19 mm when a load is suspended from it. The energy stored in the wire is then 4.8 mJ. Calculate:
(a) the mass of the load
(b) the Young modulus for the steel.

(a) Energy stored $= \frac{1}{2} F \Delta L$

  $4.8 \times 10^{-3} = \frac{1}{2} F \times (0.19 \times 10^{-3})$

  $F = 51 \text{ N}$

  Mass of the load $= \dfrac{F}{g} = \dfrac{51}{9.8} = 5.2 \text{ kg}$

(b) Stress $= \dfrac{51}{\pi \times (1 \times 10^{-3})^2} = 1.6 \times 10^7 \text{ Pa}$

  Strain $= \dfrac{0.19 \times 10^{-3}}{2.5} = 7.6 \times 10^{-5}$

  Young modulus $= \dfrac{1.6 \times 10^7}{7.6 \times 10^{-5}} = 2.1 \times 10^{11} \text{ Pa}$

## CALCULATING ENERGY STORED FROM STRESS–STRAIN GRAPHS

- Area under the initial linear part of a stress–strain graph is energy stored per unit volume.

- **Area under graph $= \frac{1}{2} \times$ stress $\times$ strain**

  $= \dfrac{1}{2} \times \dfrac{F}{A} \times \dfrac{\Delta L}{L}$

  $= \dfrac{\text{energy stored}}{\text{volume of wire}}$

- If stress is in Pa, then energy stored is in J m$^{-3}$

# STRETCHING MATERIALS

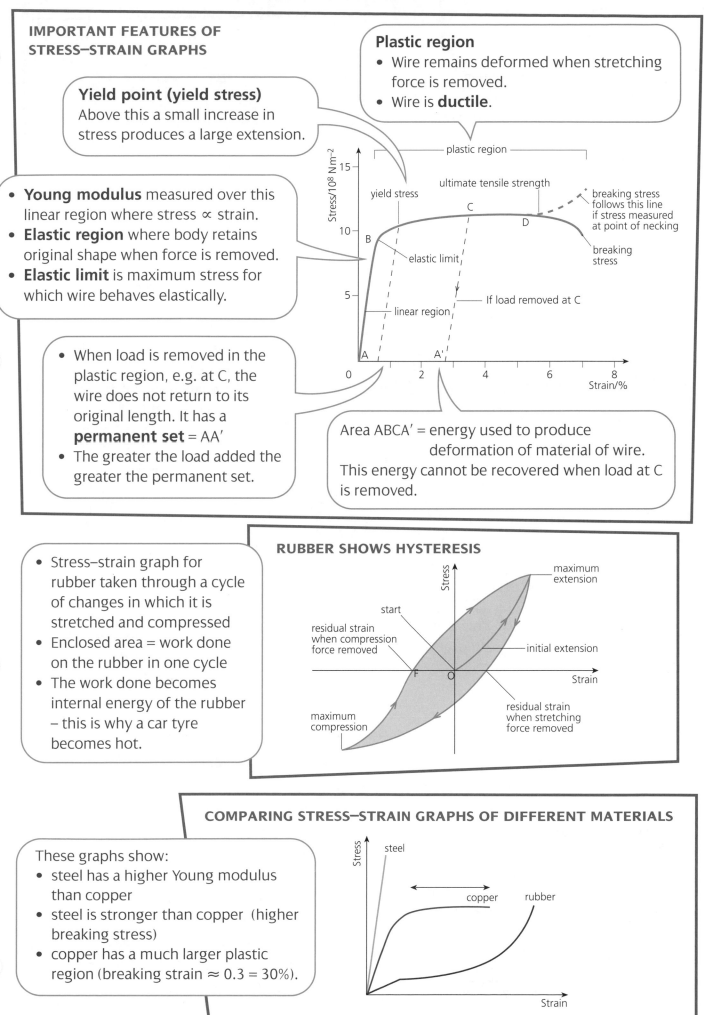

**IMPORTANT FEATURES OF STRESS–STRAIN GRAPHS**

**Yield point (yield stress)**
Above this a small increase in stress produces a large extension.

- **Young modulus** measured over this linear region where stress ∝ strain.
- **Elastic region** where body retains original shape when force is removed.
- **Elastic limit** is maximum stress for which wire behaves elastically.

**Plastic region**
- Wire remains deformed when stretching force is removed.
- Wire is **ductile**.

- When load is removed in the plastic region, e.g. at C, the wire does not return to its original length. It has a **permanent set** = AA′
- The greater the load added the greater the permanent set.

Area ABCA′ = energy used to produce deformation of material of wire. This energy cannot be recovered when load at C is removed.

## RUBBER SHOWS HYSTERESIS

- Stress–strain graph for rubber taken through a cycle of changes in which it is stretched and compressed
- Enclosed area = work done on the rubber in one cycle
- The work done becomes internal energy of the rubber – this is why a car tyre becomes hot.

## COMPARING STRESS–STRAIN GRAPHS OF DIFFERENT MATERIALS

These graphs show:
- steel has a higher Young modulus than copper
- steel is stronger than copper (higher breaking stress)
- copper has a much larger plastic region (breaking strain ≈ 0.3 = 30%).

# MICROSCOPIC BEHAVIOUR OF MATERIALS

## EXPLAINING THE BEHAVIOUR OF METALS

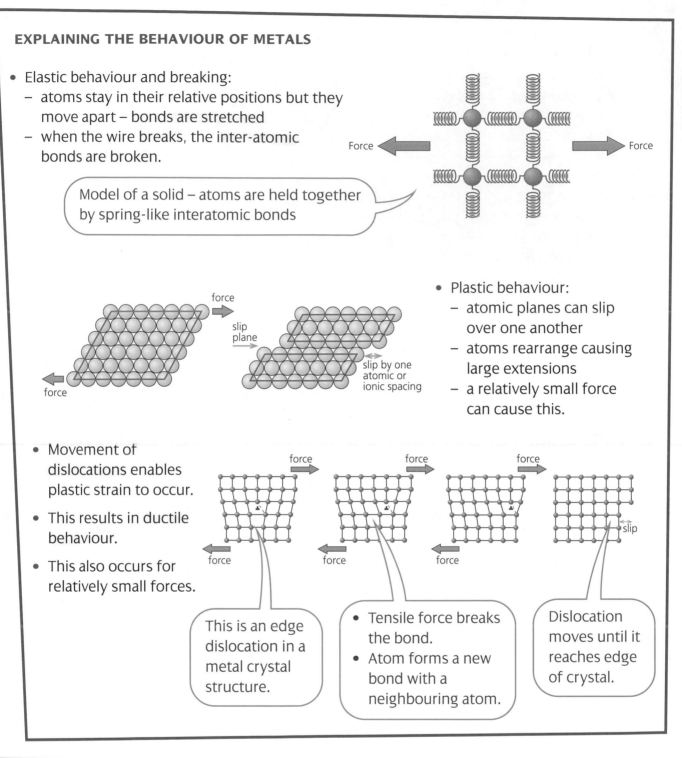

- Elastic behaviour and breaking:
  - atoms stay in their relative positions but they move apart – bonds are stretched
  - when the wire breaks, the inter-atomic bonds are broken.

Model of a solid – atoms are held together by spring-like interatomic bonds

force

slip plane

slip by one atomic or ionic spacing

force

- Plastic behaviour:
  - atomic planes can slip over one another
  - atoms rearrange causing large extensions
  - a relatively small force can cause this.

- Movement of dislocations enables plastic strain to occur.
- This results in ductile behaviour.
- This also occurs for relatively small forces.

force        force        force

force        force        force

slip

This is an edge dislocation in a metal crystal structure.

- Tensile force breaks the bond.
- Atom forms a new bond with a neighbouring atom.

Dislocation moves until it reaches edge of crystal.

## EXPLAINING THE BEHAVIOUR OF RUBBER

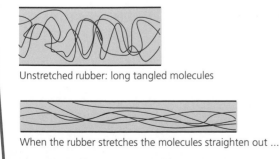

Unstretched rubber: long tangled molecules

When the rubber stretches the molecules straighten out …

… until they can't get any longer without breaking

- Rubber consists of tangled molecular chains.
- Hard to stretch at first – tangled molecular chains align.
- Then easier to stretch – long chain molecules straighten out.
- Then harder to stretch again – interatomic bonds are stretched.
- Finally, bonds break.

# GRAVITATIONAL FIELDS

- A body has mass and this mass can influence any other mass wherever it is in the Universe. This influence is called **gravity**. The mass is said to produce a **gravitational field**.

- The existence of a gravitational field is shown by an object experiencing a force due to its mass.

## NEWTON'S GRAVITATIONAL LAW

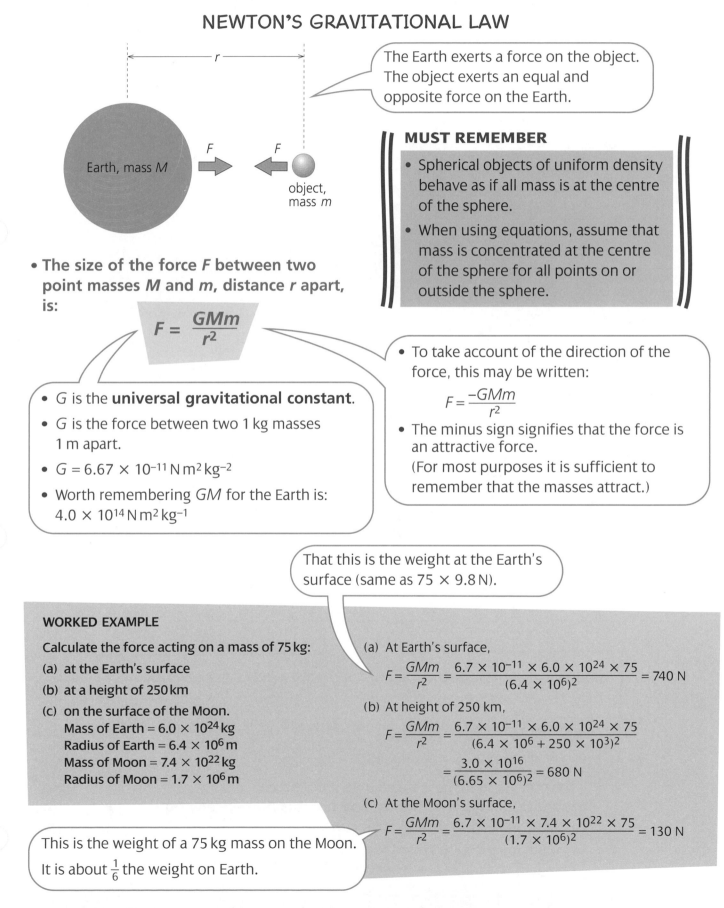

The Earth exerts a force on the object. The object exerts an equal and opposite force on the Earth.

**MUST REMEMBER**

- Spherical objects of uniform density behave as if all mass is at the centre of the sphere.

- When using equations, assume that mass is concentrated at the centre of the sphere for all points on or outside the sphere.

- **The size of the force $F$ between two point masses $M$ and $m$, distance $r$ apart, is:**

$$F = \frac{GMm}{r^2}$$

- $G$ is the **universal gravitational constant**.

- $G$ is the force between two 1 kg masses 1 m apart.

- $G = 6.67 \times 10^{-11} \, \text{N m}^2 \, \text{kg}^{-2}$

- Worth remembering $GM$ for the Earth is: $4.0 \times 10^{14} \, \text{N m}^2 \, \text{kg}^{-1}$

- To take account of the direction of the force, this may be written:

$$F = \frac{-GMm}{r^2}$$

- The minus sign signifies that the force is an attractive force.
  (For most purposes it is sufficient to remember that the masses attract.)

That this is the weight at the Earth's surface (same as $75 \times 9.8$ N).

**WORKED EXAMPLE**

Calculate the force acting on a mass of 75 kg:

(a) at the Earth's surface

(b) at a height of 250 km

(c) on the surface of the Moon.
   Mass of Earth = $6.0 \times 10^{24}$ kg
   Radius of Earth = $6.4 \times 10^6$ m
   Mass of Moon = $7.4 \times 10^{22}$ kg
   Radius of Moon = $1.7 \times 10^6$ m

(a) At Earth's surface,
$$F = \frac{GMm}{r^2} = \frac{6.7 \times 10^{-11} \times 6.0 \times 10^{24} \times 75}{(6.4 \times 10^6)^2} = 740 \text{ N}$$

(b) At height of 250 km,
$$F = \frac{GMm}{r^2} = \frac{6.7 \times 10^{-11} \times 6.0 \times 10^{24} \times 75}{(6.4 \times 10^6 + 250 \times 10^3)^2}$$
$$= \frac{3.0 \times 10^{16}}{(6.65 \times 10^6)^2} = 680 \text{ N}$$

(c) At the Moon's surface,
$$F = \frac{GMm}{r^2} = \frac{6.7 \times 10^{-11} \times 7.4 \times 10^{22} \times 75}{(1.7 \times 10^6)^2} = 130 \text{ N}$$

This is the weight of a 75 kg mass on the Moon. It is about $\frac{1}{6}$ the weight on Earth.

# GRAVITATIONAL FIELD STRENGTH

- **The gravitational field strength is the force acting on a mass of 1 kg at a point in a gravitational field.**

  > It is more correct to write this as the force acting **per** kilogram.

- Gravitational field strength is a vector quantity.
- Putting $m = 1$ kg in Newton's gravitational equation, gravitational field strength of the Earth, $F_E = \dfrac{GM}{r^2}$
- Unit of gravitational field strength is **N kg$^{-1}$**
- At the Earth's surface, $F_E$ is **9.8 N kg$^{-1}$** toward the centre of the Earth.

> **MUST REMEMBER**
>
> **Total gravitational force** on an object = gravitational field strength × mass of the object
> This is the object's **weight**.

## WORKED EXAMPLE

The distance between the Earth and the Moon is $3.8 \times 10^8$ m. Use the data on the previous page to calculate the distance from Earth at which the gravitational field strengths are equal and opposite.

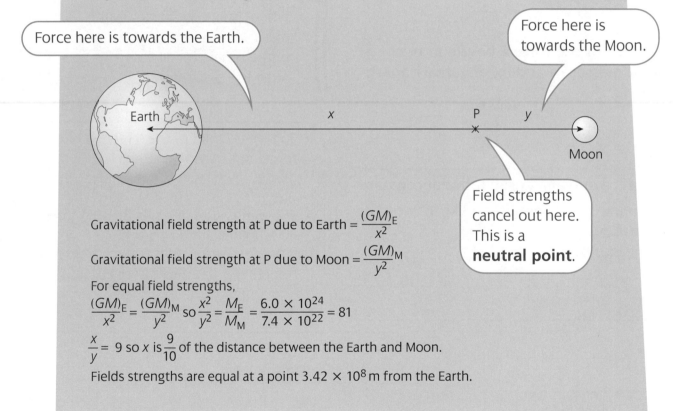

Force here is towards the Earth.

Force here is towards the Moon.

Field strengths cancel out here. This is a **neutral point**.

Gravitational field strength at P due to Earth $= \dfrac{(GM)_E}{x^2}$

Gravitational field strength at P due to Moon $= \dfrac{(GM)_M}{y^2}$

For equal field strengths,

$\dfrac{(GM)_E}{x^2} = \dfrac{(GM)_M}{y^2}$ so $\dfrac{x^2}{y^2} = \dfrac{M_E}{M_M} = \dfrac{6.0 \times 10^{24}}{7.4 \times 10^{22}} = 81$

$\dfrac{x}{y} = 9$ so $x$ is $\dfrac{9}{10}$ of the distance between the Earth and Moon.

Fields strengths are equal at a point $3.42 \times 10^8$ m from the Earth.

---

### LINKING GRAVITATIONAL FIELD STRENGTH AND ACCELERATION DUE TO GRAVITY $g$

- Gravitational force acting on an object of mass $m = \dfrac{GMm}{r^2}$
- Acceleration produced by the force $= \dfrac{F}{m} = \dfrac{GM}{r^2} = g$
- If the only force acting is that due to gravity then: **gravitational acceleration at any point is numerically equal to the gravitational field strength.**

  $$9.8 \text{ N kg}^{-1} \equiv 9.8 \text{ m s}^{-2}$$

# RADIAL AND UNIFORM FIELDS

## RADIAL FIELD

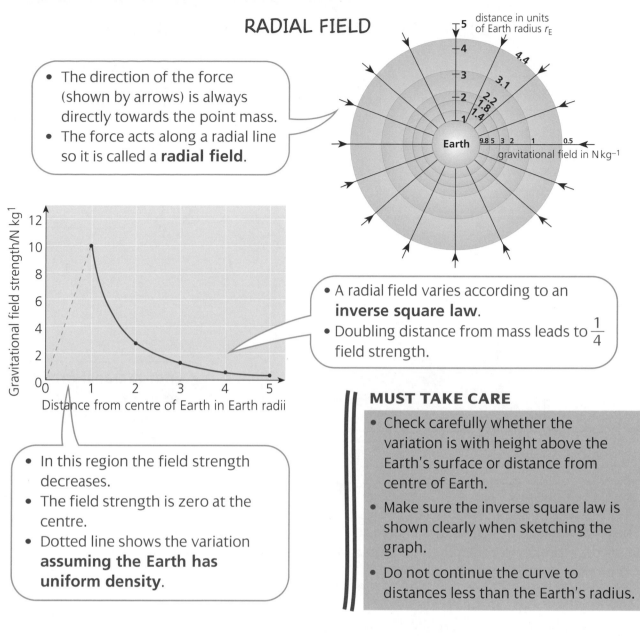

- The direction of the force (shown by arrows) is always directly towards the point mass.
- The force acts along a radial line so it is called a **radial field**.

- A radial field varies according to an **inverse square law**.
- Doubling distance from mass leads to $\frac{1}{4}$ field strength.

- In this region the field strength decreases.
- The field strength is zero at the centre.
- Dotted line shows the variation **assuming the Earth has uniform density**.

### MUST TAKE CARE

- Check carefully whether the variation is with height above the Earth's surface or distance from centre of Earth.
- Make sure the inverse square law is shown clearly when sketching the graph.
- Do not continue the curve to distances less than the Earth's radius.

## UNIFORM FIELD

- In a **uniform field** the gravitational field strength is the same at all points.
- The field near the Earth's surface is approximately uniform for a small area on the surface and for small changes in height.
- The force is then in the same direction and equal to $10\,\text{N}\,\text{kg}^{-1}$.

### WHAT IS WEIGHT?

- Weight is the force exerted on an object due to the gravitational force $= mg$
- **Weight = mass × gravitational field strength**
- Only at infinity will an object have no weight.

### MUST TAKE CARE

**Apparent weightlessness** can be experienced in orbit (see 'Gravitational potential energy', page 60).

Weightlessness is often associated with no reaction force on a surface, but the downward force due to gravitation (weight) still exists in orbit.

# GRAVITATIONAL POTENTIAL ENERGY

## POTENTIAL ENERGY CHANGES IN A UNIFORM FIELD

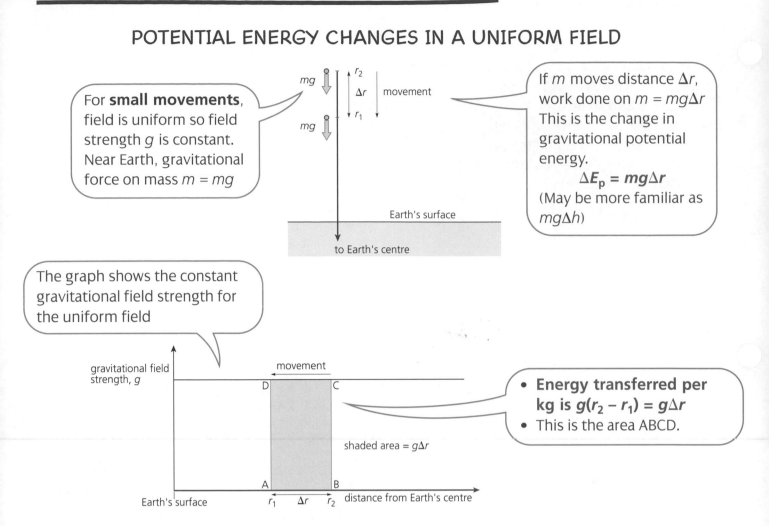

For **small movements**, field is uniform so field strength $g$ is constant. Near Earth, gravitational force on mass $m = mg$

If $m$ moves distance $\Delta r$, work done on $m = mg\Delta r$ This is the change in gravitational potential energy.
$$\Delta E_p = mg\Delta r$$
(May be more familiar as $mg\Delta h$)

The graph shows the constant gravitational field strength for the uniform field

- **Energy transferred per kg is $g(r_2 - r_1) = g\Delta r$**
- This is the area ABCD.

shaded area = $g\Delta r$

- The area between the gravitational field strength graph line and the distance axis gives the work done per kg (or energy transferred per kg) as the object moves from one point to another.

## POTENTIAL ENERGY CHANGES FOR LARGE MOVEMENT

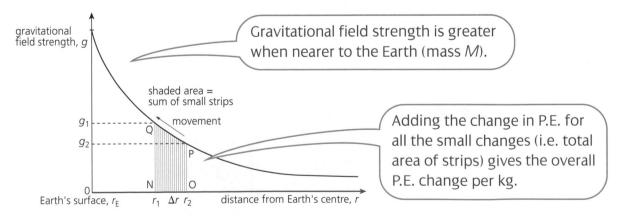

Gravitational field strength is greater when nearer to the Earth (mass $M$).

shaded area = sum of small strips

Adding the change in P.E. for all the small changes (i.e. total area of strips) gives the overall P.E. change per kg.

- The change in potential energy $\Delta E_p$ as a mass $m$ moves from $r_2$ to $r_1$ can be found using calculus. This gives:

$$\Delta E_p = GMm\left(\frac{1}{r_2} - \frac{1}{r_1}\right)$$

- For movement toward the Earth, this is a decrease in potential energy.
- If falling freely, the mass gains this much kinetic energy.

# ABSOLUTE GRAVITATIONAL POTENTIAL

- **Gravitational potential at a point P in a gravitational field is the change in potential energy of 1 kg when it moves from infinity to P.**

- Gravitational potential is measured in **J kg⁻¹**

- An object at infinity has no force acting on it so has no potential energy.

- The potential at infinity is zero.

- As an object moves from infinity it loses potential energy so gravitational potentials are negative.

$$\text{Gravitational potential} = GM\left(\frac{1}{\infty} - \frac{1}{r}\right) = \frac{-GM}{r}$$

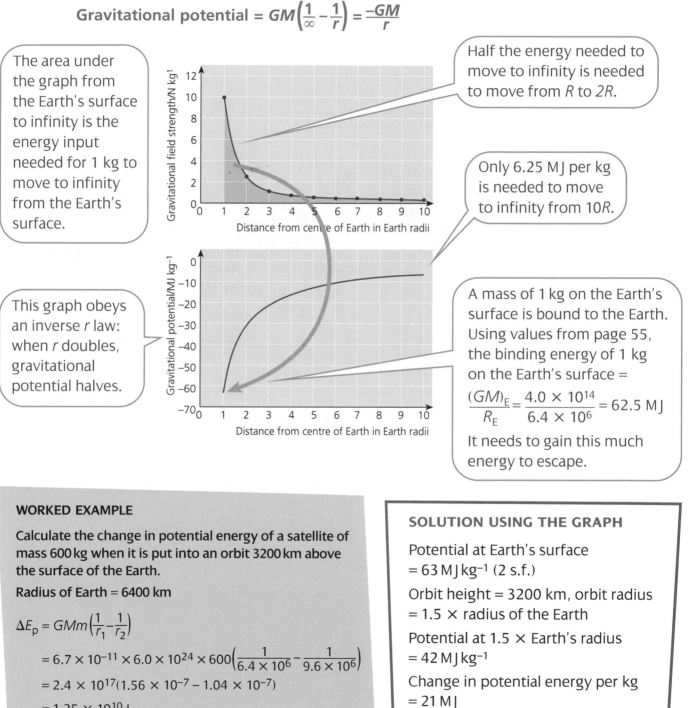

The area under the graph from the Earth's surface to infinity is the energy input needed for 1 kg to move to infinity from the Earth's surface.

Half the energy needed to move to infinity is needed to move from $R$ to $2R$.

Only 6.25 MJ per kg is needed to move to infinity from $10R$.

This graph obeys an inverse $r$ law: when $r$ doubles, gravitational potential halves.

A mass of 1 kg on the Earth's surface is bound to the Earth. Using values from page 55, the binding energy of 1 kg on the Earth's surface =
$$\frac{(GM)_E}{R_E} = \frac{4.0 \times 10^{14}}{6.4 \times 10^6} = 62.5 \text{ MJ}$$
It needs to gain this much energy to escape.

---

**WORKED EXAMPLE**

Calculate the change in potential energy of a satellite of mass 600 kg when it is put into an orbit 3200 km above the surface of the Earth.

Radius of Earth = 6400 km

$$\Delta E_p = GMm\left(\frac{1}{r_1} - \frac{1}{r_2}\right)$$

$$= 6.7 \times 10^{-11} \times 6.0 \times 10^{24} \times 600\left(\frac{1}{6.4 \times 10^6} - \frac{1}{9.6 \times 10^6}\right)$$

$$= 2.4 \times 10^{17}(1.56 \times 10^{-7} - 1.04 \times 10^{-7})$$

$$= 1.25 \times 10^{10} \text{ J}$$

**SOLUTION USING THE GRAPH**

Potential at Earth's surface
= 63 MJ kg⁻¹ (2 s.f.)

Orbit height = 3200 km, orbit radius
= 1.5 × radius of the Earth

Potential at 1.5 × Earth's radius
= 42 MJ kg⁻¹

Change in potential energy per kg
= 21 MJ

Total P.E. change for 600 kg
= 1.26 × 10¹⁰ J

---

**MUST TAKE CARE**

- Use distances from centre of Earth in the formula (**not** heights).
- Add radius of Earth to heights.

# ORBITS

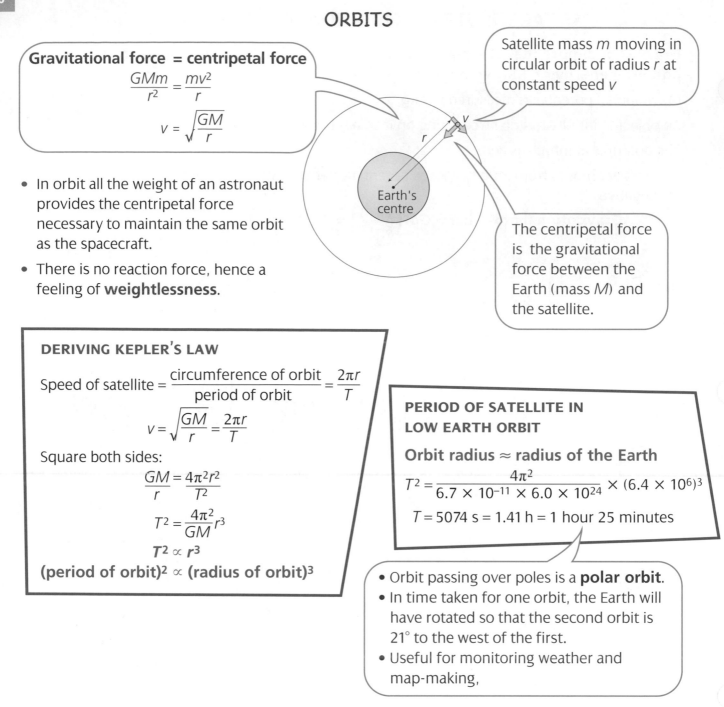

**Gravitational force = centripetal force**

$$\frac{GMm}{r^2} = \frac{mv^2}{r}$$

$$v = \sqrt{\frac{GM}{r}}$$

Satellite mass $m$ moving in circular orbit of radius $r$ at constant speed $v$

- In orbit all the weight of an astronaut provides the centripetal force necessary to maintain the same orbit as the spacecraft.

- There is no reaction force, hence a feeling of **weightlessness**.

r

Earth's centre

v

The centripetal force is the gravitational force between the Earth (mass $M$) and the satellite.

### DERIVING KEPLER'S LAW

$$\text{Speed of satellite} = \frac{\text{circumference of orbit}}{\text{period of orbit}} = \frac{2\pi r}{T}$$

$$v = \sqrt{\frac{GM}{r}} = \frac{2\pi r}{T}$$

Square both sides:

$$\frac{GM}{r} = \frac{4\pi^2 r^2}{T^2}$$

$$T^2 = \frac{4\pi^2}{GM} r^3$$

$$T^2 \propto r^3$$

**(period of orbit)² ∝ (radius of orbit)³**

### PERIOD OF SATELLITE IN LOW EARTH ORBIT

**Orbit radius ≈ radius of the Earth**

$$T^2 = \frac{4\pi^2}{6.7 \times 10^{-11} \times 6.0 \times 10^{24}} \times (6.4 \times 10^6)^3$$

$$T = 5074 \text{ s} = 1.41 \text{ h} = 1 \text{ hour } 25 \text{ minutes}$$

- Orbit passing over poles is a **polar orbit**.
- In time taken for one orbit, the Earth will have rotated so that the second orbit is 21° to the west of the first.
- Useful for monitoring weather and map-making,

# GEOSYNCHRONOUS (GEOSTATIONARY) ORBIT

- This is a west-to-east orbit around the equator.
- Orbit period is 24 hours so a satellite stays above the same point on the equator.
- Orbit used for communication satellites.
- Receivers can always point to the same place in space – no tracking necessary.

### RADIUS OF ORBIT OF GEOSYNCHRONOUS SATELLITE

$$T^2 = \frac{4\pi^2}{GM} r^3$$

$$(24 \times 60 \times 60)^2 = \frac{4\pi^2}{4.0 \times 10^{14}} r^3$$

$$r = 4.23 \times 10^7 \text{ m}$$

# TOTAL ENERGY IN ORBIT

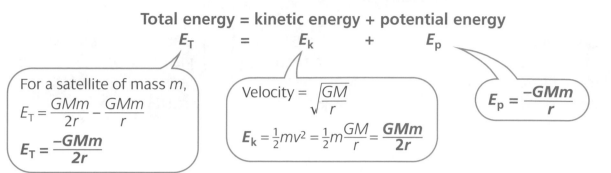

**Total energy = kinetic energy + potential energy**

$$E_T \quad = \quad E_k \quad + \quad E_p$$

For a satellite of mass $m$,
$$E_T = \frac{GMm}{2r} - \frac{GMm}{r}$$
$$E_T = \frac{-GMm}{2r}$$

Velocity $= \sqrt{\dfrac{GM}{r}}$
$$E_k = \tfrac{1}{2}mv^2 = \tfrac{1}{2}m\frac{GM}{r} = \frac{GMm}{2r}$$

$$E_p = \frac{-GMm}{r}$$

---

**WORKED EXAMPLE**

Calculate the minimum energy required to put a 100 kg satellite into a geosynchronous orbit from an equatorial launch site.
$GM$ for Earth $= 4.0 \times 10^{14}\,\text{N m}^2\,\text{kg}^{-1}$
Radius of geosynchronous orbit $= 4.2 \times 10^7\,\text{m}$

K.E. of 1 kg at the equator $= \tfrac{1}{2}v^2$
$$= \tfrac{1}{2}\left(\frac{2\pi R_E}{24 \times 60 \times 60}\right)^2$$
$$= 0.11\,\text{MJ}$$

P.E. of 1 kg at equator $= \dfrac{-GMm}{R_E}$
$$= \frac{-4.0 \times 10^{14} \times 1}{6400 \times 10^3} = -62.5\,\text{MJ}$$

Total energy $= -62.4\,\text{MJ}$

Total energy of 1 kg mass in orbit $= \dfrac{-GMm}{2r}$
$$= \frac{4.0 \times 10^{14}}{2 \times 4.2 \times 10^7}$$
$$= -4.8\,\text{MJ}$$

Energy input needed per kg $= -4.8 - (-62.4) = 57.6\,\text{MJ}$

Total energy needed to put 100 kg into orbit
$$= 57.6 \times 100 = 5760\,\text{MJ}$$

**MUST REMEMBER**

The negative sign for total energy shows that this amount of energy has to be supplied for a satellite to escape.

---

## ESCAPE VELOCITY

- This is the velocity necessary for a body to escape from the surface of an astronomical object.

- To escape requires an input of energy equal to the gravitational potential for each kg.
  **Initial kinetic energy needed = change in potential energy**

$$\tfrac{1}{2}mv^2 = \frac{GMm}{R} \text{ where } R \text{ is radius and } M \text{ mass of the astronomical object.}$$

**Escape velocity, $v = \sqrt{\dfrac{2GM}{R}}$**

---

**WORKED EXAMPLES**

1 Calculate the escape velocity from the surface of the Earth.
2 For a black hole, the escape velocity must be at least the speed of light $(3.0 \times 10^8\,\text{m s}^{-1})$. To what radius would the Sun have to shrink for it to become a black hole? (Assume that classical laws hold.)
Mass of Sun $= 2.0 \times 10^{30}\,\text{kg}$
What would the density of matter be in the Sun then?

1 $v = \sqrt{\dfrac{2GM}{R}} = \sqrt{\dfrac{2 \times 4.0 \times 10^{14}}{6400 \times 10^3}} = 11.2\,\text{km s}^{-1}$

2 $3.0 \times 10^8 = \sqrt{\dfrac{2 \times 6.7 \times 10^{-11} \times 2.0 \times 10^{30}}{R}}$

Radius at which Sun would become a black hole is 2980 m

Density $= \dfrac{\text{mass}}{\text{volume}} = \dfrac{2.0 \times 10^{30}}{\frac{4}{3}\pi(2980)^3}$
$$= 1.8 \times 10^{19}\,\text{kg m}^{-3}$$

---

- Calculations of escape velocity make the unrealistic assumption that all the K.E. needed is provided instantaneously. The high acceleration required is impossible and the effects of the atmosphere on the motion of the payload (producing retardation and a temperature rise) also have to be considered.

## FINDING MASSES OF EARTH AND SUN

$G$ measured in a laboratory experiment (Cavendish's experiment):
$6.7 \times 10^{-11} \, N \, m^2 \, kg^{-2}$

Radius $R$ of Earth measured from known curvature:
$6400 \, km$

$g$ at the Earth's surface measured using free fall or pendulum experiment:
$9.8 \, N \, kg^{-1}$

Mass of Earth $M_E$ calculated from
$$g = \frac{GM_E}{R^2}:$$
$6.0 \times 10^{24} \, kg$

Period $T_E$ of orbit of Earth around Sun is measured (1 year):
$3.15 \times 10^7 \, s$

Radius $r_E$ of Earth's orbit around Sun measured:
$1.5 \times 10^{11} \, m$

Mass of Sun $M_S$ calculated from
$$T_E{}^2 = \frac{4\pi^2}{GM_S} r_E{}^3:$$
$2.0 \times 10^{30} \, kg$

## WHAT IS THE FUTURE OF THE UNIVERSE?

- Start of Universe
- Big Bang
- Matter given kinetic energy
- Matter moves in all directions

- Matter moves away
- P.E. ($E_p$) increases
- Each bit of matter attracted by the rest

- **Total energy = $E_k + E_p$**
- $E_k$ is always positive and falls as $E_p$ increases (Note $E_p$ is negative as $E_p$ = zero at infinity.)

- If there is sufficient matter in the Universe, then K.E. falls to zero at a finite radius.
  – Matter moves back to origin of the Big Bang leading to a **Big Crunch**.

- In the critical condition (flat Universe) there is sufficient matter, so that eventually (after an infinite time) everything just stops moving.
  – Matter was just given sufficient energy to escape at the Big Bang.

- If there is insufficient matter, the Universe expands more quickly than in the critical condition and continues to expand forever.

# CHARGE AND ELECTRIC FIELDS

- Like charges repel each other.
- Unlike charges attract each other.
- A charged body produces an electric field.
- In electric fields charged objects feel a force.
- The electric field direction is the direction of the force on a positively charged object.

**ELECTRIC FIELD STRENGTH, E**

- **This is the force acting on one coulomb of charge (more precisely, force per unit charge).**
- The unit of $E$ is **N C$^{-1}$**

$$E = \frac{F}{Q}$$

- For a charge $Q$ in a field of strength $E$, $F = EQ$

## REPRESENTING FIELDS

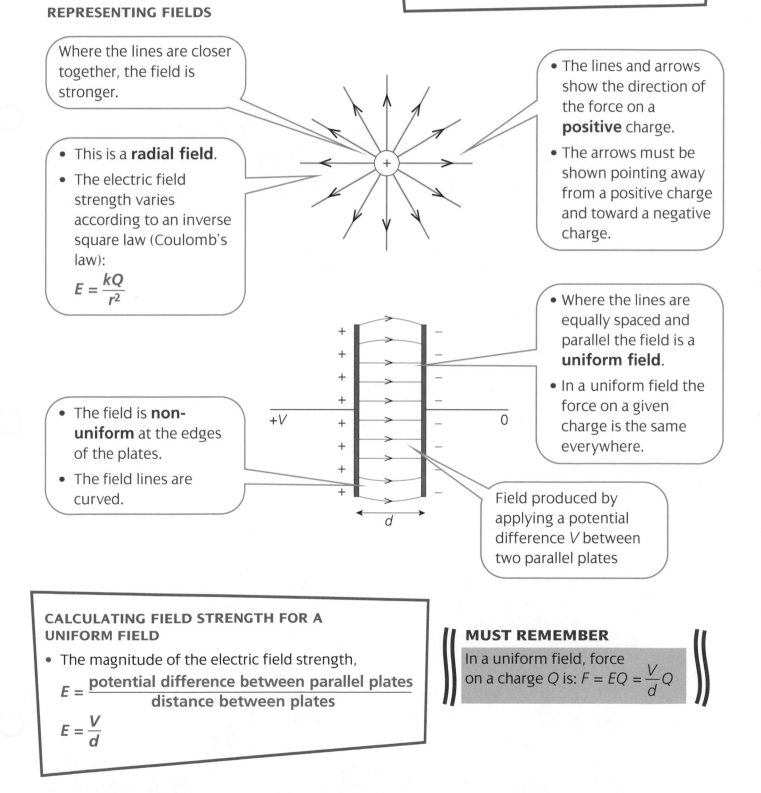

Where the lines are closer together, the field is stronger.

- This is a **radial field**.
- The electric field strength varies according to an inverse square law (Coulomb's law):

$$E = \frac{kQ}{r^2}$$

- The lines and arrows show the direction of the force on a **positive** charge.
- The arrows must be shown pointing away from a positive charge and toward a negative charge.

- The field is **non-uniform** at the edges of the plates.
- The field lines are curved.

- Where the lines are equally spaced and parallel the field is a **uniform field**.
- In a uniform field the force on a given charge is the same everywhere.

Field produced by applying a potential difference $V$ between two parallel plates

## CALCULATING FIELD STRENGTH FOR A UNIFORM FIELD

- The magnitude of the electric field strength,

$$E = \frac{\text{potential difference between parallel plates}}{\text{distance between plates}}$$

$$E = \frac{V}{d}$$

**MUST REMEMBER**

In a uniform field, force on a charge $Q$ is: $F = EQ = \dfrac{V}{d}Q$

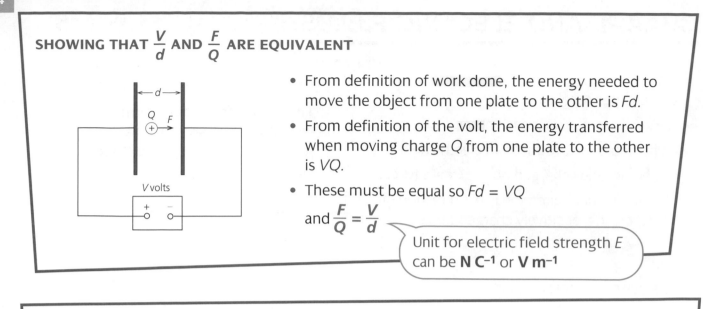

## SHOWING THAT $\frac{V}{d}$ AND $\frac{F}{Q}$ ARE EQUIVALENT

- From definition of work done, the energy needed to move the object from one plate to the other is $Fd$.
- From definition of the volt, the energy transferred when moving charge $Q$ from one plate to the other is $VQ$.
- These must be equal so $Fd = VQ$

and $\frac{F}{Q} = \frac{V}{d}$

Unit for electric field strength $E$ can be **N C⁻¹** or **V m⁻¹**

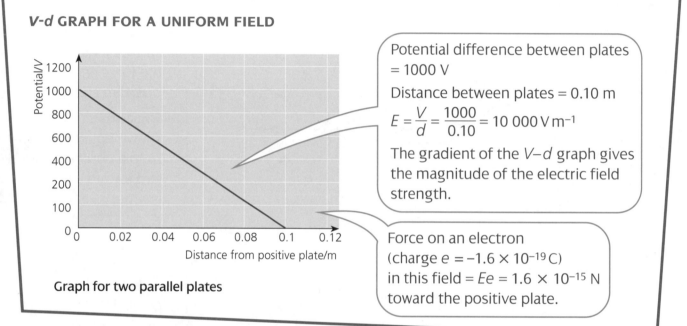

## V-d GRAPH FOR A UNIFORM FIELD

Graph for two parallel plates

Potential difference between plates = 1000 V

Distance between plates = 0.10 m

$E = \frac{V}{d} = \frac{1000}{0.10} = 10\,000\,\text{V m}^{-1}$

The gradient of the $V-d$ graph gives the magnitude of the electric field strength.

Force on an electron (charge $e = -1.6 \times 10^{-19}$ C) in this field $= Ee = 1.6 \times 10^{-15}$ N toward the positive plate.

### WORKED EXAMPLE (RADIAL FIELD)

1 Calculate the magnitude of the electric field strength 15 cm from a charge of 3.5 pC.

$E = \frac{kQ}{r^2}$

$\quad = \frac{9 \times 10^9 \times 3.5 \times 10^{-12}}{0.15^2}$

$\quad = 1.4\,\text{N C}^{-1}$

2 Calculate the field strength 45 cm from the charge of 3.5 pC.

The distance has increased by 3

so $E$ is $\frac{1}{9}$ of $1.4 = 0.16\,\text{N C}^{-1}$

or

$\frac{E_1}{E_2} = \frac{r_1^2}{r_2^2}$ so $\frac{E_2}{1.4} = \frac{15^2}{45^2}$

$\qquad E_2 = 0.16\,\text{N C}^{-1}$

### WORKED EXAMPLE (UNIFORM FIELD)

1 Calculate the field strength between two parallel plates that are separated by 0.45 m when a potential difference of 150 V is applied across the plates.

Field strength between the plates, $E = \frac{V}{d}$

$E = \frac{150}{0.45}$

$\quad = 330\,\text{V m}^{-1}\,(\text{N C}^{-1})$

2 Calculate the force on an alpha particle (charge $+3.2 \times 10^{-19}$ C) when it is between the plates.

Force on an alpha particle between the plates

$F = EQ$

$F = 330 \times 3.2 \times 10^{-19}$

$F = 1.1 \times 10^{-16}$ N

# ELECTROSTATIC FORCE BETWEEN POINT CHARGES

## COULOMB'S LAW

- This gives force between charges $Q_1$ and $Q_2$ a distance $r$ apart in a vacuum.

- $F = \dfrac{Q_1 Q_2}{4\pi\varepsilon_0 r^2}$

  where $\varepsilon_0$ is the **permittivity** of free space $\varepsilon_0 = 8.9 \times 10^{-12}\,\text{F m}^{-1}$

- If medium between charges is not a vacuum,

  $$F = \dfrac{Q_1 Q_2}{4\pi\varepsilon_0\varepsilon_r r^2}$$

  where $\varepsilon_r$ is the **relative permittivity** of the medium.

**Repulsion**
Two like charges

Two opposite charges
**Attraction**

- Permittivity of medium $\varepsilon = \varepsilon_0\varepsilon_r$
- The relative permittivity of air $\approx 1$

### MUST REMEMBER

- There is a force on both charges.
- The forces on the charges are equal and opposite.

### MUST TAKE CARE

$F$ is in newton, N if:
$Q$ is in coulomb, C
$r$ is in metre, m.

### MAGNITUDE OF ELECTRIC FIELD STRENGTH IN A RADIAL FIELD

- **Electric field strength $E$ at a point is the force on a charge of +1 C at that point.**
  (**Force per unit charge** is more correct.)

- Field strength due to a charge $Q$ in air or vacuum,

  $$E = \dfrac{Q}{4\pi\varepsilon_0 r^2} = \dfrac{kQ}{r^2}$$

$k = \dfrac{1}{4\pi\varepsilon_0} = 8.9 \times 10^9\,\text{F}^{-1}\text{m}$
$\approx 9 \times 10^9\,\text{F}^{-1}\text{m}$ (or $\text{N m}^2\text{C}^{-2}$)

- This shows variation of $E$ with $r$ for a point charge or for a sphere of radius 0.01 m.
- A charged sphere behaves as if the charge is at its centre, for points outside it.
- The graph is asymptotic to the $E$- and $d$-axes (i.e. it **must not intersect either axis**).

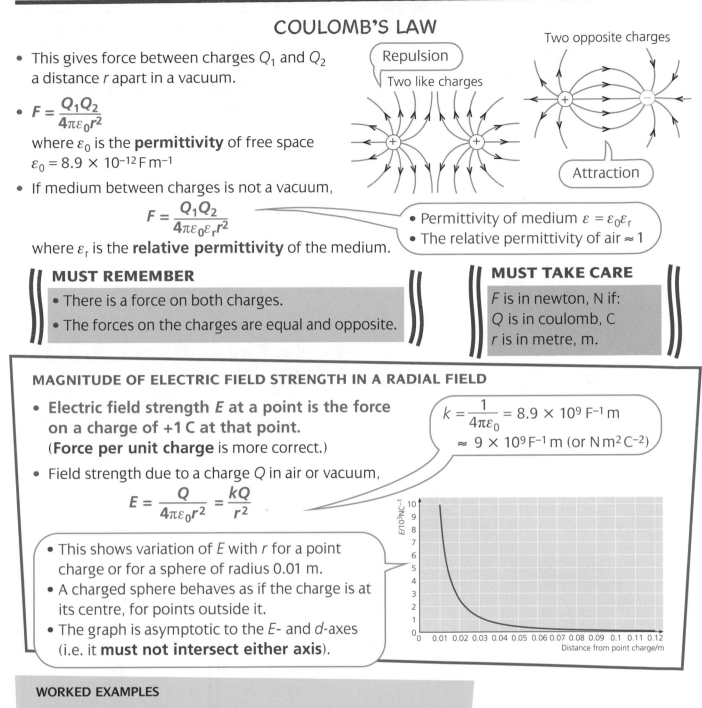

$E/10^3\text{NC}^{-1}$ vs Distance from point charge/m

### WORKED EXAMPLES

1 Calculate the magnitude of the point charge that gives rise to the above graph with air as the medium.

At distance of 0.010 m, $E = 10\,000\ \text{N C}^{-1}$

$E = \dfrac{Q}{4\pi\varepsilon_0 r^2} = \dfrac{kQ}{r^2}$

$10\,000 = \dfrac{8.9 \times 10^9\,Q}{0.010^2}$

$Q = 1.12 \times 10^{-10}\,\text{C} = 112\,\text{pC}$

### MUST TAKE CARE

Must be careful with the squared quantities when using a calculator.

2 Calculate the force between a positive and negative ion each with charge $1.6 \times 10^{-19}\,\text{C}$ which are $6.0 \times 10^{-10}\,\text{m}$ apart in water of relative permittivity 80.

$F = \dfrac{Q_1 Q_2}{4\pi\varepsilon_0\varepsilon_r r^2}$

$= \dfrac{(1.6 \times 10^{-19})^2}{4 \times \pi \times 8.9 \times 10^{-12} \times 80 \times (6.0 \times 10^{-10})^2}$

$= 7.9 \times 10^{-12}\,\text{N}$

# ELECTRIC POTENTIAL

- Electric potential at infinity is defined as zero (0 V).
- **Electrical potential at a point is the energy required move +1 C of charge from infinity to that point.**

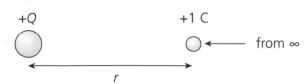

+Q            +1 C
              ← from ∞

r

- This is often called the **absolute potential**.
- More precisely, the potential is the **energy per coulomb** needed to move charge from infinity to the point.
- Unit of electrical potential is **J C⁻¹**

- Work is done to move the +1 C charge against the repulsive force between the charges.
- So for a positive charge, energy has to be provided – the potential is positive.
- The potential $V$ is the total work done moving the charge from infinity to distance $r$ from +Q:

$$V = \frac{+Q}{4\pi\varepsilon_0 r}$$

- A negative charge attracts a +1 C charge, so no energy input is needed to move it from infinity.
- A negative charge loses potential energy when brought from infinity, so the potential is negative.
- The potential at a point due to a negative charge is given by: $V = \frac{-Q}{4\pi\varepsilon_0 r}$
- When released, a negative charge will move away (to infinity) gaining kinetic energy.

- $E$ at any point is minus the gradient of $V$–$r$ graph.

$$E = \frac{-\Delta V}{\Delta r}$$

- This is minus the potential gradient.

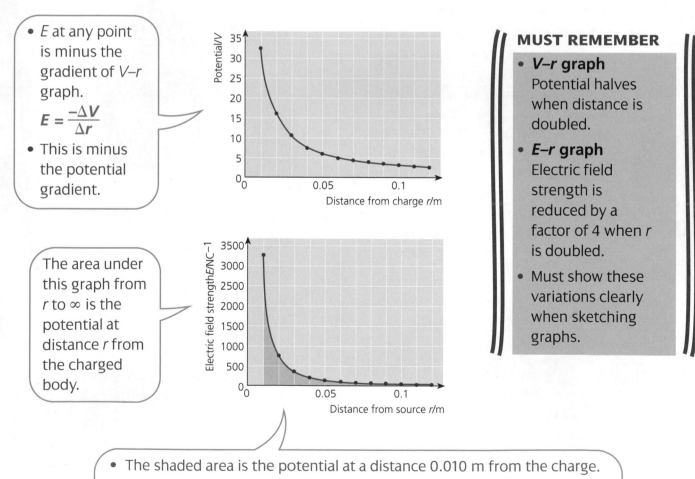

- **$V$–$r$ graph**
  Potential halves when distance is doubled.
- **$E$–$r$ graph**
  Electric field strength is reduced by a factor of 4 when $r$ is doubled.
- Must show these variations clearly when sketching graphs.

The area under this graph from $r$ to ∞ is the potential at distance $r$ from the charged body.

- The shaded area is the potential at a distance 0.010 m from the charge.
- **To find an approximate value of electrical potential:**
  – estimate the number of squares under the graph line (≈ 6 squares)
  – multiply by the number of J C⁻¹ represented by one square (5 J C⁻¹).

# POTENTIAL DIFFERENCE IN A RADIAL ELECTRIC FIELD

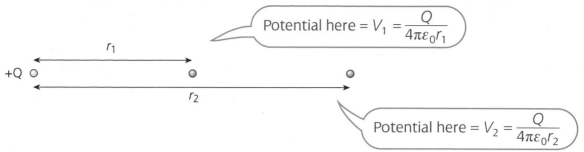

Potential here $= V_1 = \dfrac{Q}{4\pi\varepsilon_0 r_1}$

Potential here $= V_2 = \dfrac{Q}{4\pi\varepsilon_0 r_2}$

- The charge is positive so $V_1 > V_2$
- **Magnitude of potential difference, $\Delta V = V_1 - V_2$**

$$\Delta V = \frac{Q}{4\pi\varepsilon_0}\left(\frac{1}{r_1} - \frac{1}{r_2}\right)$$

- A positive charge loses P.E. when it moves from $r_1$ to $r_2$

---

**CHANGE IN ELECTRICAL POTENTIAL ENERGY**

- **Change in potential energy = potential difference × charge moved**

$$\Delta E_p = \Delta V \times q$$

- **Magnitude of change in potential energy $\Delta E_p = \dfrac{Qq}{4\pi\varepsilon_0}\left(\dfrac{1}{r_1} - \dfrac{1}{r_2}\right)$**

---

**MUST TAKE CARE**

- If $Q$ and $q$ are both positive, or both negative, then there is a **decrease** in P.E. as they separate.
- If $Q$ and $q$ have opposite charges, the P.E. **increases** as they separate.

If troubled by signs:
- use equation to find magnitude
- then consider signs and positions of charges to decide whether P.E. increases or decreases.

---

**WORKED EXAMPLE**

**(a)** Calculate the potential at a distance of 8.0 cm from a point charge of −25 pC.

**(b)** Calculate the change in potential energy when a charge of +1.5 pC moves from this point to a point that is 12.0 cm from the −25 pC charge.

(a) Potential $= \dfrac{Q}{4\pi\varepsilon_0 r} = \dfrac{-25 \times 10^{-12}}{4 \times 3.14 \times 8.9 \times 10^{-12} \times 0.08}$

$= -2.8$ V

(b) Change in P.E. = final P.E. − initial P.E.

$\Delta E_p = \dfrac{Qq}{4\pi\varepsilon_0}\left(\dfrac{1}{r_1} - \dfrac{1}{r_2}\right)$

$= \dfrac{-25 \times 10^{-12} \times 1.5 \times 10^{-12}}{4 \times 3.14 \times 8.9 \times 10^{-12}}\left(\dfrac{1}{0.12} - \dfrac{1}{0.08}\right)$

$= -3.35 \times 10^{-13}(8.33 - 12.5) = +1.4 \times 10^{-12}$ J

The answer is positive so there is an increase in P.E.

The positive and negative charges attract, so that work is done on the charges to separate them and the P.E. must increase.

# EQUIPOTENTIALS

- **Equipotential lines** are like contour lines (lines of equal height above sea level) on a map.
- The potential at all points on an equipotential line or surface is the same at all points on it.

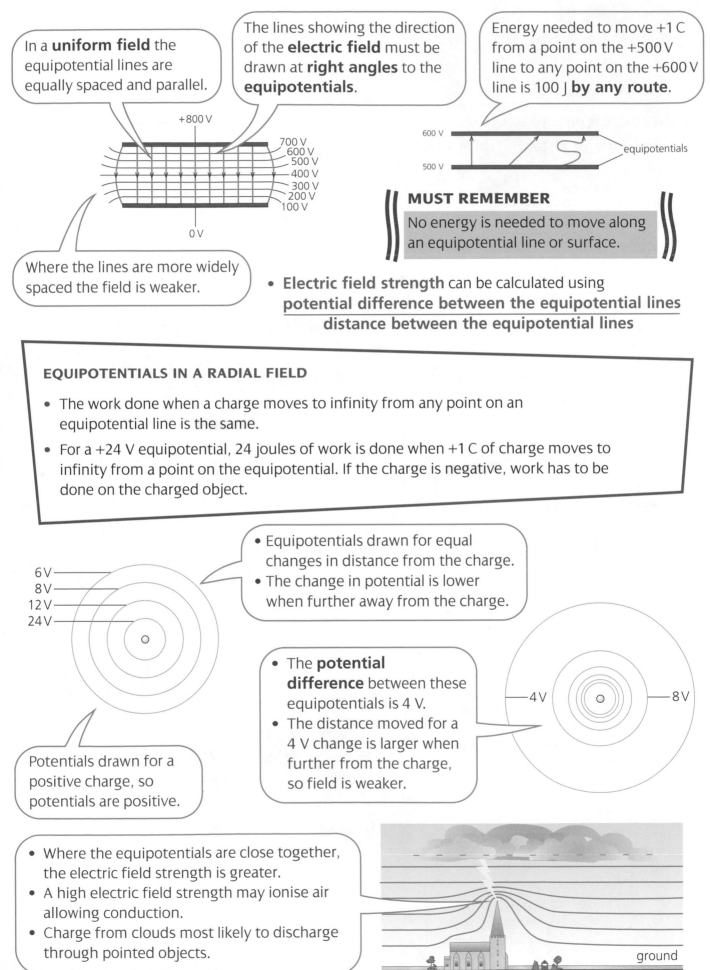

In a **uniform field** the equipotential lines are equally spaced and parallel.

The lines showing the direction of the **electric field** must be drawn at **right angles** to the **equipotentials**.

Energy needed to move +1 C from a point on the +500 V line to any point on the +600 V line is 100 J **by any route**.

Where the lines are more widely spaced the field is weaker.

## MUST REMEMBER

No energy is needed to move along an equipotential line or surface.

- **Electric field strength** can be calculated using

$$\frac{\text{potential difference between the equipotential lines}}{\text{distance between the equipotential lines}}$$

## EQUIPOTENTIALS IN A RADIAL FIELD

- The work done when a charge moves to infinity from any point on an equipotential line is the same.
- For a +24 V equipotential, 24 joules of work is done when +1 C of charge moves to infinity from a point on the equipotential. If the charge is negative, work has to be done on the charged object.

- Equipotentials drawn for equal changes in distance from the charge.
- The change in potential is lower when further away from the charge.

Potentials drawn for a positive charge, so potentials are positive.

- The **potential difference** between these equipotentials is 4 V.
- The distance moved for a 4 V change is larger when further from the charge, so field is weaker.

- Where the equipotentials are close together, the electric field strength is greater.
- A high electric field strength may ionise air allowing conduction.
- Charge from clouds most likely to discharge through pointed objects.

# FORCE AND POTENTIAL DUE TO CHARGED SPHERES

- For points outside a charged sphere the charge behaves as if it is at the centre of the sphere.

- Inside a charged sphere:
  - The potential is the same as the potential at the surface.
  - There is no potential gradient, so no electric field.

**WORKED EXAMPLE**

The sphere of a van de Graaff generator has a radius of 0.20 m. The potential of the sphere is 20 000 V. Calculate:
(a) the charge on the sphere
(b) the capacitance of the sphere
(c) the field strength at the surface
(d) the distance from the surface at which the potential falls to 5000 V.

(a)
$$V = \frac{Q}{4\pi\varepsilon_0 r}$$
$$20\,000 = \frac{Q}{4\pi \times 8.9 \times 10^{-12} \times 0.2}$$
$$Q = 0.45\,\mu C$$

(b) Capacitance $= \dfrac{Q}{V} = \dfrac{4.5 \times 10^{-7}}{20\,000} = 22\,\text{pF}$
(See 'Capacitors', page 73)

(c) $E = \dfrac{Q}{4\pi\varepsilon_0 r^2} = 100\,000\,\text{V m}^{-1}$

(d) $V$ reduced to $\frac{1}{4}$ so distance from centre is multiplied by 4.
Potential falls to 5000 V at 0.80 m from the centre of the sphere.
Distance from surface = 0.60 m

**MUST TAKE CARE**

Read carefully whether distance required is from surface or centre of sphere.

# COMPARING ELECTRIC AND GRAVITATIONAL FIELDS

- The forces for point charges and masses vary according to an inverse square $(\frac{1}{r^2})$ law.
- The potentials vary according to a $\frac{1}{r}$ law.

- Gravitational fields only produce attraction.
- Electric fields may produce attraction or repulsion.
- Electrostatic force depends on medium between the charges.

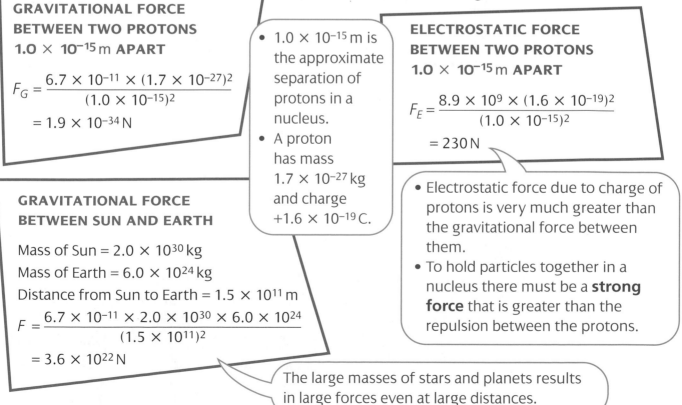

**GRAVITATIONAL FORCE BETWEEN TWO PROTONS 1.0 × 10⁻¹⁵ m APART**

$$F_G = \frac{6.7 \times 10^{-11} \times (1.7 \times 10^{-27})^2}{(1.0 \times 10^{-15})^2}$$
$$= 1.9 \times 10^{-34}\,\text{N}$$

- $1.0 \times 10^{-15}$ m is the approximate separation of protons in a nucleus.
- A proton has mass $1.7 \times 10^{-27}$ kg and charge $+1.6 \times 10^{-19}$ C.

**ELECTROSTATIC FORCE BETWEEN TWO PROTONS 1.0 × 10⁻¹⁵ m APART**

$$F_E = \frac{8.9 \times 10^9 \times (1.6 \times 10^{-19})^2}{(1.0 \times 10^{-15})^2}$$
$$= 230\,\text{N}$$

- Electrostatic force due to charge of protons is very much greater than the gravitational force between them.
- To hold particles together in a nucleus there must be a **strong force** that is greater than the repulsion between the protons.

**GRAVITATIONAL FORCE BETWEEN SUN AND EARTH**

Mass of Sun = $2.0 \times 10^{30}$ kg
Mass of Earth = $6.0 \times 10^{24}$ kg
Distance from Sun to Earth = $1.5 \times 10^{11}$ m
$$F = \frac{6.7 \times 10^{-11} \times 2.0 \times 10^{30} \times 6.0 \times 10^{24}}{(1.5 \times 10^{11})^2}$$
$$= 3.6 \times 10^{22}\,\text{N}$$

The large masses of stars and planets results in large forces even at large distances.

# DEFLECTING PARTICLES

## USING GRAVITATIONAL FIELDS

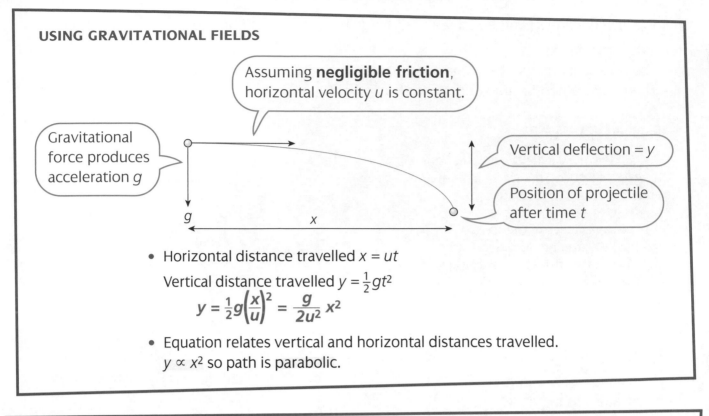

Assuming **negligible friction**, horizontal velocity $u$ is constant.

Gravitational force produces acceleration $g$

Vertical deflection = $y$

Position of projectile after time $t$

$g$

$x$

- Horizontal distance travelled $x = ut$

  Vertical distance travelled $y = \frac{1}{2}gt^2$

  $$y = \frac{1}{2}g\left(\frac{x}{u}\right)^2 = \frac{g}{2u^2}x^2$$

- Equation relates vertical and horizontal distances travelled.
  $y \propto x^2$ so path is parabolic.

## USING AN ELECTRIC FIELD

- Vertical deflection while between plates,

  $$s = \frac{1}{2}at^2$$
  $$= \frac{1}{2}\frac{Eq}{m}\left(\frac{L}{u}\right)^2$$

- $E$, $q$, $m$ and $u$ are all constants, so $s \propto L^2$
- Deflection proportional to square of distance travelled horizontally.
- **Path between plates is parabolic**.

- Horizontal velocity $u$ is constant.
- Time between plates $t = \dfrac{L}{u}$
- $u$ is usually very high so $t$ is short and deflection due to gravity is negligible.

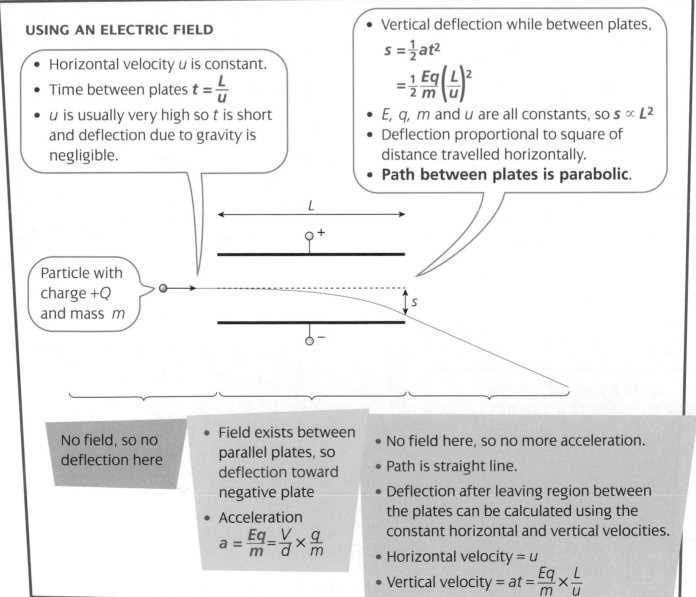

$L$

$+$

Particle with charge $+Q$ and mass $m$

$s$

$-$

No field, so no deflection here

- Field exists between parallel plates, so deflection toward negative plate
- Acceleration
  $$a = \frac{Eq}{m} = \frac{V}{d} \times \frac{q}{m}$$

- No field here, so no more acceleration.
- Path is straight line.
- Deflection after leaving region between the plates can be calculated using the constant horizontal and vertical velocities.
- Horizontal velocity = $u$
- Vertical velocity = $at = \dfrac{Eq}{m} \times \dfrac{L}{u}$

# USING ELECTRIC FIELDS

## MILLIKAN'S EXPERIMENT

- To find the charge on an electron, Millikan balanced the forces produced by an electric field and gravitational attraction on a charged oil drop.

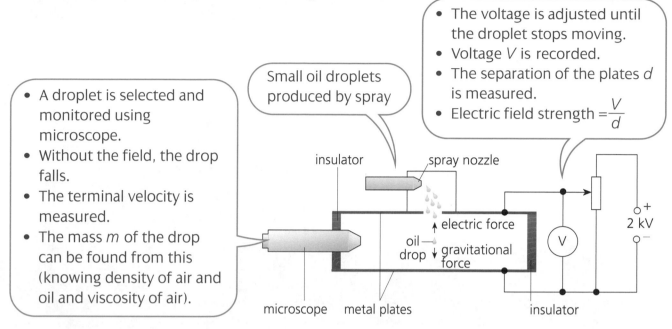

- The voltage is adjusted until the droplet stops moving.
- Voltage $V$ is recorded.
- The separation of the plates $d$ is measured.
- Electric field strength $= \dfrac{V}{d}$

Small oil droplets produced by spray

- A droplet is selected and monitored using microscope.
- Without the field, the drop falls.
- The terminal velocity is measured.
- The mass $m$ of the drop can be found from this (knowing density of air and oil and viscosity of air).

insulator   spray nozzle

electric force

oil drop   gravitational force

microscope   metal plates   insulator

2 kV

- When fields balance, weight = force due to electric field
  $mg = Eq$ where $q$ is the charge on the drop

- Millikan measured charges on many drops – all were multiples of $-1.6 \times 10^{-19}$ C so he concluded this to be the electron charge.

## ACCELERATING ELECTRONS

- The electron gun is used to accelerate charges in oscilloscopes and television tubes.

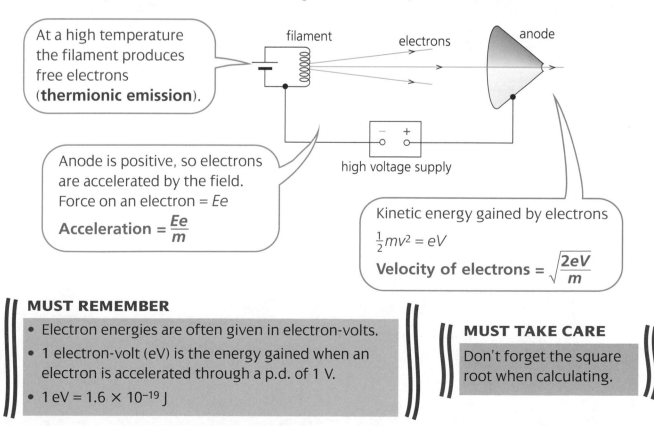

At a high temperature the filament produces free electrons (**thermionic emission**).

filament   electrons   anode

high voltage supply

Anode is positive, so electrons are accelerated by the field.
Force on an electron $= Ee$

**Acceleration** $= \dfrac{Ee}{m}$

Kinetic energy gained by electrons
$\frac{1}{2}mv^2 = eV$

**Velocity of electrons** $= \sqrt{\dfrac{2eV}{m}}$

**MUST REMEMBER**

- Electron energies are often given in electron-volts.
- 1 electron-volt (eV) is the energy gained when an electron is accelerated through a p.d. of 1 V.
- $1\,eV = 1.6 \times 10^{-19}$ J

**MUST TAKE CARE**

Don't forget the square root when calculating.

# FINDING CLOSEST DISTANCE OF APPROACH

- When a particle (e.g. proton, deuteron or alpha particle) is fired directly at a nucleus:
  – if the energy is low, it will stop and return along the original path
  – if energy is high enough, it can combine with the nucleus to produce a nuclear transformation.

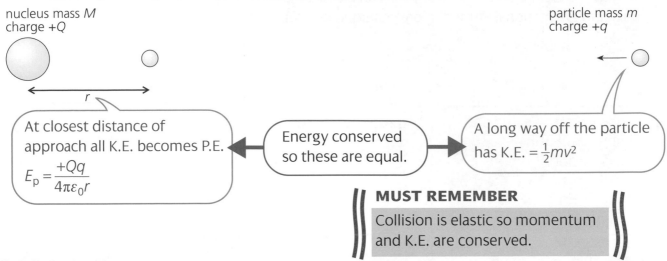

nucleus mass $M$
charge $+Q$

particle mass $m$
charge $+q$

$r$

At closest distance of approach all K.E. becomes P.E.
$$E_p = \frac{+Qq}{4\pi\varepsilon_0 r}$$

Energy conserved so these are equal.

A long way off the particle has K.E. $= \frac{1}{2}mv^2$

**MUST REMEMBER**

Collision is elastic so momentum and K.E. are conserved.

**WORKED EXAMPLE**

Calculate:
(a) the closest distance of approach of a 4.5 MeV alpha particle (charge $2e$) to a gold nucleus (charge $79e$)
(b) the minimum energy needed for the alpha particle to come into contact with the nucleus.
alpha particle radius $= 2.0 \times 10^{-15}$ m, gold nucleus radius $= 7.0 \times 10^{-15}$ m, $e = 1.6 \times 10^{-19}$ C

(a) At closest distance, $E_p =$ initial K.E.

Conversion of 4.5 MeV to J
(1 eV $= 1.6 \times 10^{-19}$ J)

$$\frac{Qq}{4\pi\varepsilon_0 r} = 4.5 \times 10^6 \times 1.6 \times 10^{-19}$$

$$\frac{(79 \times 1.6 \times 10^{-19}) \times (2 \times 1.6 \times 10^{-19})}{4\pi \times 8.9 \times 10^{-12} r} = 7.2 \times 10^{-13}$$

Closest distance, $r = 5.0 \times 10^{-14}$ m

(b) Energy needed $= \dfrac{(79 \times 1.6 \times 10^{-19}) \times (2 \times 1.6 \times 10^{-19})}{4\pi \times 8.9 \times 10^{-12} \times 9.0 \times 10^{-15}}$

$= 4.0 \times 10^{-12}$ J $= 25$ MeV

They touch when closest distance of approach is equal to sum of radii.

**MINIMUM TEMPERATURE REQUIRED FOR NUCLEAR FUSION**

- Typical nuclei involved are tritium and deuterium both having radius of about $1.3 \times 10^{-15}$ m. They must touch so that the strong force causes fusion.

- Energy needed to touch $= \dfrac{(1.6 \times 10^{-19}) \times (1.6 \times 10^{-19})}{4\pi \times 8.9 \times 10^{-12} \times 2.6 \times 10^{-15}} = 8.8 \times 10^{-14}$ J

- Mean K.E. of nuclei $= \frac{3}{2}kT = 4.4 \times 10^{-14}$ J

Each nucleus must have half the total energy needed.

($k$ is Boltzmann's constant $1.4 \times 10^{-23}$ J K$^{-1}$
Mean temperature $= 2.1 \times 10^9$ K)

**MUST REMEMBER**

- Nuclei have range of energies.
- Some nuclei have sufficient energy to fuse at much lower temperatures.

# CAPACITORS

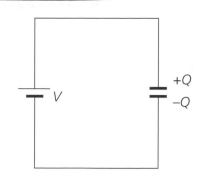

- A capacitor is charged by connecting a voltage $V$ between the plates.
- When the capacitor charge is $Q$, there is a charge of $+Q$ on one plate and $-Q$ on the other.
- **The capacitance $C$ is the charge per volt:**

$$C = \frac{Q}{V} \text{ or } Q = CV$$

- The unit of capacitance is farad F.
- 1 farad = 1 coulomb per volt

## MUST TAKE CARE

Must not confuse the symbol for capacitance with the unit for charge:
- capacitance $C$ is in farad F
- charge $Q$ is in coulomb C

## MUST REMEMBER

- The shaded area under the graph gives the energy stored by the capacitor. (It is really equal to the area between the line and the $Q$-axis as energy is the area under a $V$–$Q$ graph.)
- The gradient of the charge against voltage graph is the capacitance of the capacitor.

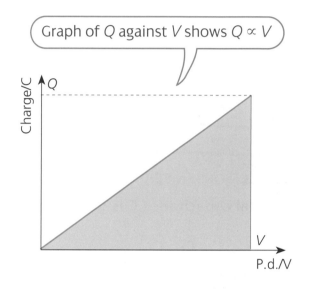

Graph of $Q$ against $V$ shows $Q \propto V$

## MEANING OF MARKINGS ON A CAPACITOR

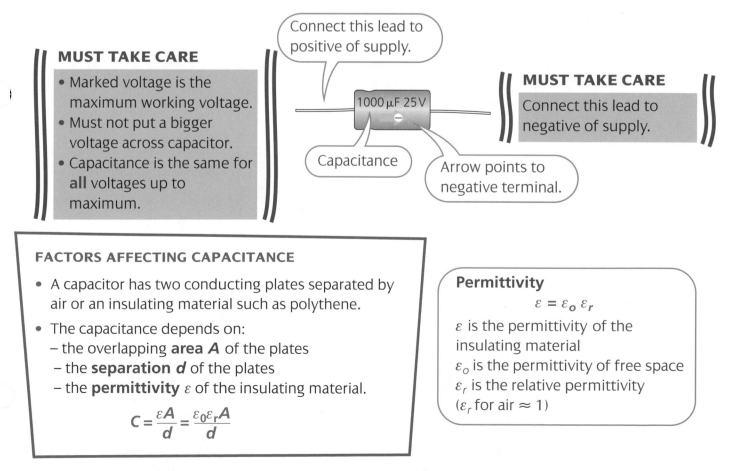

Connect this lead to positive of supply.

1000 μF 25 V

## MUST TAKE CARE

- Marked voltage is the maximum working voltage.
- Must not put a bigger voltage across capacitor.
- Capacitance is the same for **all** voltages up to maximum.

## MUST TAKE CARE

Connect this lead to negative of supply.

Capacitance

Arrow points to negative terminal.

### FACTORS AFFECTING CAPACITANCE

- A capacitor has two conducting plates separated by air or an insulating material such as polythene.
- The capacitance depends on:
  – the overlapping **area $A$** of the plates
  – the **separation $d$** of the plates
  – the **permittivity $\varepsilon$** of the insulating material.

$$C = \frac{\varepsilon A}{d} = \frac{\varepsilon_0 \varepsilon_r A}{d}$$

**Permittivity**

$$\varepsilon = \varepsilon_0 \varepsilon_r$$

$\varepsilon$ is the permittivity of the insulating material
$\varepsilon_0$ is the permittivity of free space
$\varepsilon_r$ is the relative permittivity
($\varepsilon_r$ for air $\approx 1$)

# COMBINING CAPACITORS

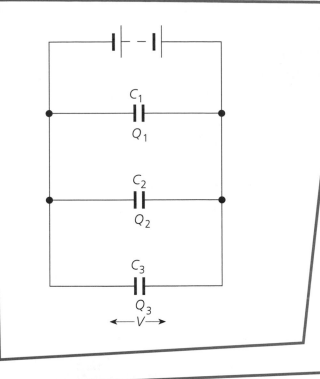

## CAPACITORS IN PARALLEL

- **Total capacitance $C = C_1 + C_2 + C_3$**
- Voltage across each capacitor is the same.
- Charge shared in ratio of capacitances.

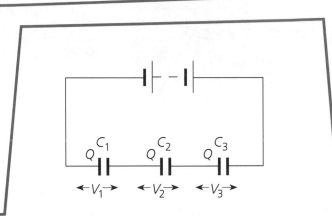

## CAPACITORS IN SERIES

- **Total capacitance $C$ is given by:**
$$\frac{1}{C} = \frac{1}{C_1} + \frac{1}{C_2} + \frac{1}{C_3}$$
- Charge on each capacitor is the same.
- Adding the voltages across the capacitors gives the supply e.m.f.

### MUST TAKE CARE

- Add voltages as a check when doing problems.
- In series, the smallest capacitor has the largest voltage.
- In parallel, the smallest capacitor has the smallest charge.

### WORKED EXAMPLE

Three capacitors, $100\,\mu F$, $200\,\mu F$ and $300\,\mu F$, are connected in series to a 5.0 V supply. Calculate the:
(a) total capacitance
(b) charge in each capacitor
(c) voltage across the $200\,\mu F$ capacitor
(d) energy stored in the $200\,\mu F$ capacitor.

(a) $\dfrac{1}{C} = \dfrac{1}{100} + \dfrac{1}{200} + \dfrac{1}{300} = 0.0100 + 0.0050 + 0.0033 = 0.0183\,\mu F^{-1}$

    $C = 54.6\,\mu F$

(b) Total charge on the combination of capacitors $= CV = 54.6 \times 5.0$
                                                              $= 273\,\mu C$

    This is also the charge on each capacitor.

(c) $V = \dfrac{Q}{C} = \dfrac{273\,\mu C}{200\,\mu F} = 1.37\,V$

(d) Energy stored $= \frac{1}{2}CV^2 = \frac{1}{2} \times 200 \times 10^{-6} \times 1.37^2 = 0.19\,mJ$

### MUST TAKE CARE

Do the final reciprocal stage!

### MUST KNOW

$1\,\mu F = 1 \times 10^{-6}\,F$
$1\,nF = 1 \times 10^{-9}\,F$
$1\,pF = 1 \times 10^{-12}\,F$

See page 76.

# DISCHARGING A CAPACITOR

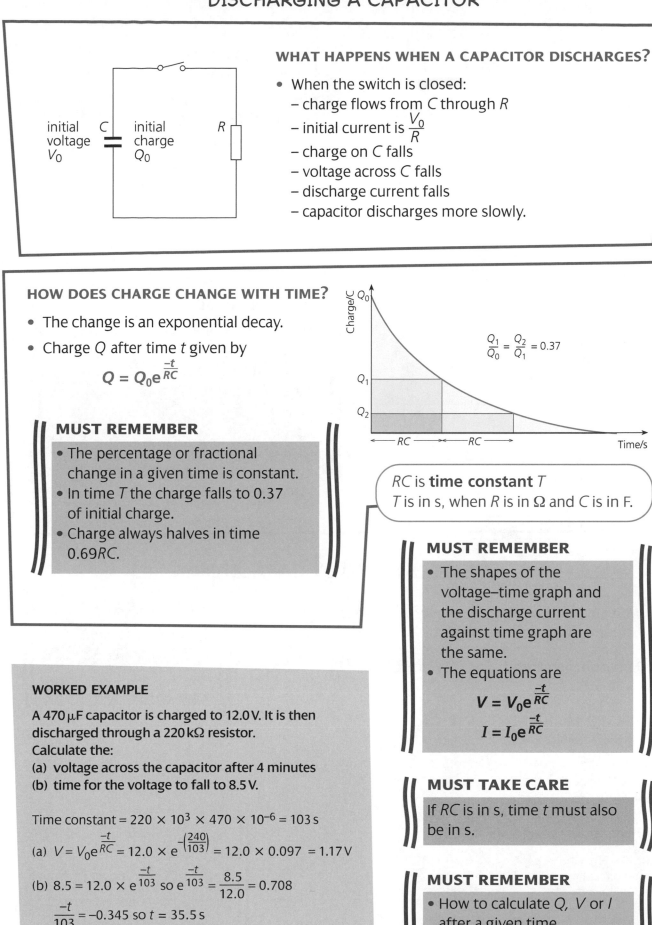

### WHAT HAPPENS WHEN A CAPACITOR DISCHARGES?

- When the switch is closed:
  - charge flows from $C$ through $R$
  - initial current is $\dfrac{V_0}{R}$
  - charge on $C$ falls
  - voltage across $C$ falls
  - discharge current falls
  - capacitor discharges more slowly.

## HOW DOES CHARGE CHANGE WITH TIME?

- The change is an exponential decay.
- Charge $Q$ after time $t$ given by

$$Q = Q_0 e^{\frac{-t}{RC}}$$

$$\frac{Q_1}{Q_0} = \frac{Q_2}{Q_1} = 0.37$$

### MUST REMEMBER

- The percentage or fractional change in a given time is constant.
- In time $T$ the charge falls to 0.37 of initial charge.
- Charge always halves in time $0.69RC$.

$RC$ is **time constant** $T$
$T$ is in s, when $R$ is in $\Omega$ and $C$ is in F.

### MUST REMEMBER

- The shapes of the voltage–time graph and the discharge current against time graph are the same.
- The equations are

$$V = V_0 e^{\frac{-t}{RC}}$$
$$I = I_0 e^{\frac{-t}{RC}}$$

### WORKED EXAMPLE

A 470 µF capacitor is charged to 12.0 V. It is then discharged through a 220 kΩ resistor.
Calculate the:
(a) voltage across the capacitor after 4 minutes
(b) time for the voltage to fall to 8.5 V.

Time constant $= 220 \times 10^3 \times 470 \times 10^{-6} = 103\,\text{s}$

(a) $V = V_0 e^{\frac{-t}{RC}} = 12.0 \times e^{-\left(\frac{240}{103}\right)} = 12.0 \times 0.097 = 1.17\,\text{V}$

(b) $8.5 = 12.0 \times e^{\frac{-t}{103}}$ so $e^{\frac{-t}{103}} = \dfrac{8.5}{12.0} = 0.708$

$\dfrac{-t}{103} = -0.345$ so $t = 35.5\,\text{s}$

### MUST TAKE CARE

If $RC$ is in s, time $t$ must also be in s.

### MUST REMEMBER

- How to calculate $Q$, $V$ or $I$ after a given time.
- How to calculate the time for these to fall to a given value or fraction of the starting value.

# CHARGING A CAPACITOR

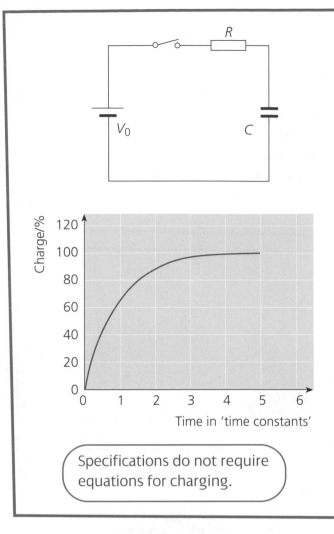

Specifications do not require equations for charging.

## WHAT HAPPENS WHEN A CAPACITOR CHARGES?

- When the switch is closed:
  - charge flows to C through R
  - initial current is $\dfrac{V_0}{R}$
  - charge on C increases
  - voltage across C increases
  - discharge current falls since p.d. across R falls
  - capacitor charges more slowly.

### MUST REMEMBER

- The charge–time and voltage–time graphs for charging have this shape.
- The charging current–time graph is the same shape as the discharge current–time graph on the previous page.
- The area under any current–time graph gives the charge that flows in that time.

# ENERGY STORED BY CAPACITORS

- When a capacitor charges, energy is transferred from the supply to the capacitor.

- **The energy stored, $E = \frac{1}{2}QV$ or $\frac{1}{2}CV^2$ or $\frac{1}{2}\dfrac{Q^2}{C}$**

### MUST TAKE CARE

$E$ is in joules
when $C$ is in farads
$V$ is in volts
and $Q$ is in coulombs.

### WORKED EXAMPLE

A 2200 µF capacitor is charged to 12.0 V and then discharged to 5.0 V through a resistor.
Calculate the:
(a) charge that flows during the discharge
(b) energy dissipated in the resistor.

(a) Initial charge $= CV = 12 \times 2200 \times 10^{-6}\,\mathrm{C} = 0.0264\,\mathrm{C}$
 Final charge $= 5.0 \times 2200 \times 10^{-6}\,\mathrm{C} = 0.0110\,\mathrm{C}$
 Charge flowing $= 0.0264 - 0.0110\,\mathrm{C} = 0.0154\,\mathrm{C}$

(b) Initial energy $= \frac{1}{2}CV^2 = \frac{1}{2}(2200 \times 10^{-6}) \times 12.0^2 = 0.158\,\mathrm{J}$
 Final energy $= \frac{1}{2}CV^2 = \frac{1}{2}(2200 \times 10^{-6}) \times 5.0^2 = 0.028$
 Energy dissipated $= 0.158 - 0.028 = 0.13\,\mathrm{J}$

### MUST TAKE CARE

- The answer to (b) is **not** the same as
$\frac{1}{2}(2200 \times 10^{-6}) \times (12.0 - 5)^2$
(This is a common mistake.)
- It is safer to calculate the initial and final energies and then subtract one from the other.

# MAGNETIC FIELDS

### WHAT IS A MAGNETIC FIELD?

- A magnetic field is a region in which a magnetic effect is observed.
- The presence of a magnetic field is shown by:
  – a force on a wire carrying a current
  – a force on a moving charged particle due to its charge
  – a force on a magnetic pole
  – an induced e.m.f. in a conductor produced by a changing magnetic field.

## FORCE ON A CURRENT-CARRYING CONDUCTOR

A force is produced when there is a magnetic field at right-angles to the current. The **force** is at **right-angles** to both **magnetic field** and **current**.

wire carrying current (conventional direction : + to –)

force on wire

N       S

magnetic field

thumb: motion force

forefinger: field

second finger: current

### MUST REMEMBER

**Fleming's left-hand rule** is used to find relative directions of magnetic field, current and force.

## MAGNETIC FLUX DENSITY $B$

- The strength of a magnetic field is defined by the **flux density $B$**.
  $B$ is measured in **tesla (T)**.

### DEFINING THE TESLA

- The magnitude of the force $F$ on a current-carrying wire increases for an increase in:
  – the flux density $B$
  – current in the wire $I$
  – length of wire in the field $l$.
- **The flux density is 1 tesla (T) when a force of 1 newton (N) acts on each metre (m) of wire when the current is 1 ampere (A).**

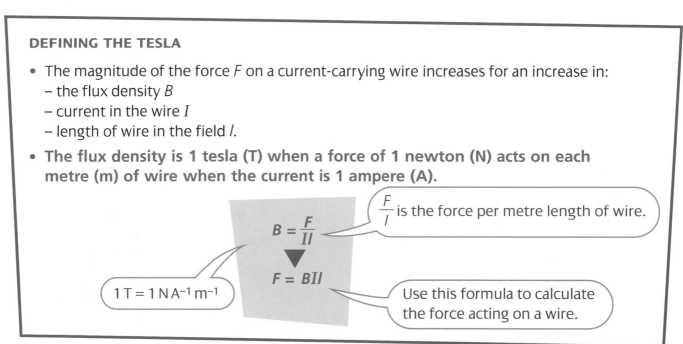

$\dfrac{F}{l}$ is the force per metre length of wire.

$$B = \frac{F}{Il}$$

▼

$$F = BIl$$

$1\,\text{T} = 1\,\text{N A}^{-1}\,\text{m}^{-1}$

Use this formula to calculate the force acting on a wire.

# PRODUCING MAGNETIC FIELDS

- Magnetic fields are produced by moving charged particles.
- Electromagnets use currents in wires to produce the field.

## FIELD DUE TO CURRENT IN A STRAIGHT LINE

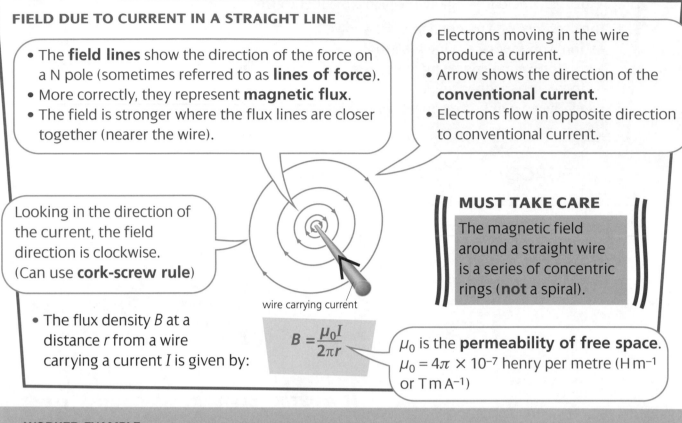

- The **field lines** show the direction of the force on a N pole (sometimes referred to as **lines of force**).
- More correctly, they represent **magnetic flux**.
- The field is stronger where the flux lines are closer together (nearer the wire).

- Electrons moving in the wire produce a current.
- Arrow shows the direction of the **conventional current**.
- Electrons flow in opposite direction to conventional current.

Looking in the direction of the current, the field direction is clockwise. (Can use **cork-screw rule**)

**MUST TAKE CARE**

The magnetic field around a straight wire is a series of concentric rings (**not** a spiral).

wire carrying current

- The flux density $B$ at a distance $r$ from a wire carrying a current $I$ is given by:

$$B = \frac{\mu_0 I}{2\pi r}$$

$\mu_0$ is the **permeability of free space**.
$\mu_0 = 4\pi \times 10^{-7}$ henry per metre (H m$^{-1}$ or T m A$^{-1}$)

### WORKED EXAMPLE

(a) Calculate the field strength at a distance of 8.0 m from a cable carrying a current of 60 A.
(b) How far from the wire would the field strength be equal to that of the horizontal component of the Earth's field (0.18 μT)?

(a) $B = \dfrac{\mu_0 I}{2\pi r} = \dfrac{4\pi \times 10^{-7} \times 60}{2\pi \times 8} = 1.5 \times 10^{-6}$ T

(b) $0.18 \times 10^{-6} = \dfrac{4\pi \times 10^{-7} \times 60}{2\pi r}$

$r = \dfrac{4\pi \times 10^{-7} \times 60}{2\pi \times 0.18 \times 10^{-6}} = 67$ m

## MAGNETIC FLUX DENSITY INSIDE A LONG SOLENOID

- The **magnetic flux density** inside an air-filled long solenoid is $\mu_0 nI$
- In practice, a long solenoid is one for which the length is at least 7× greater than the diameter.

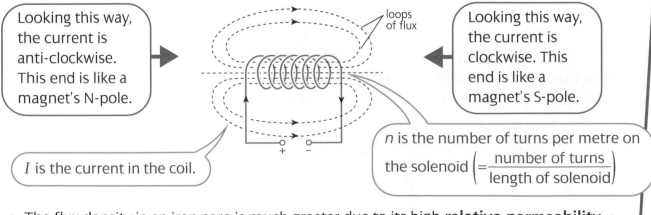

Looking this way, the current is anti-clockwise. This end is like a magnet's N-pole.

loops of flux

Looking this way, the current is clockwise. This end is like a magnet's S-pole.

$I$ is the current in the coil.

$n$ is the number of turns per metre on the solenoid $\left( = \dfrac{\text{number of turns}}{\text{length of solenoid}} \right)$

- The flux density in an iron core is much greater due to its high **relative permeability** $\mu_r$
- $\mu_r$ for iron can be up to 1000 so the magnetic flux is 1000 times greater than in air.
- For an iron-cored coil, $B = \mu_0 \mu_r nI$

# MEASURING AND INVESTIGATING FIELDS

## CURRENT BALANCE

- This is a device for measuring the force on a wire carrying a current in a magnetic field.
- If force, current and length of the wire in the field are known, the **flux density** of the field can be found.

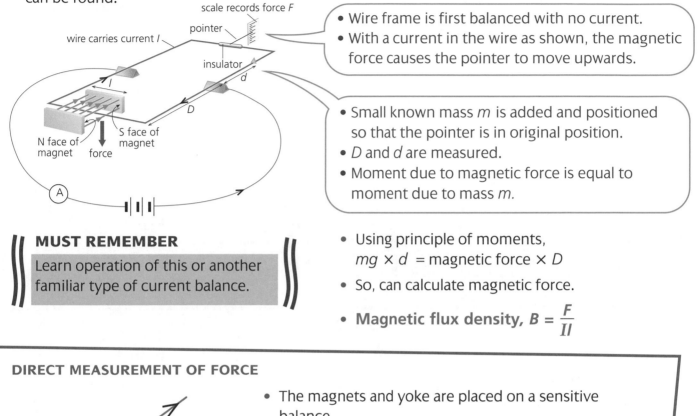

scale records force F

pointer

wire carries current I

insulator

d

D

I

N face of magnet   force   S face of magnet

A

- Wire frame is first balanced with no current.
- With a current in the wire as shown, the magnetic force causes the pointer to move upwards.

- Small known mass m is added and positioned so that the pointer is in original position.
- D and d are measured.
- Moment due to magnetic force is equal to moment due to mass m.

**MUST REMEMBER**

Learn operation of this or another familiar type of current balance.

- Using principle of moments,
  $mg \times d$ = magnetic force $\times D$
- So, can calculate magnetic force.
- **Magnetic flux density, $B = \dfrac{F}{Il}$**

---

### DIRECT MEASUREMENT OF FORCE

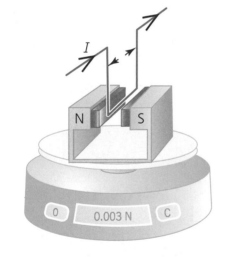

I

N   S

0   0.003 N   C

- The magnets and yoke are placed on a sensitive balance.
- The mass is measured and the downward force (= mg) recorded.
- The current is turned on and the new force recorded.
- The difference gives the force due to the field.

- Experiments can be performed to investigate how:
  – force varies with length for constant current
  – force varies with current for constant length.
- If length and current are measured, then can find magnetic flux density.

---

### INVESTIGATING FLUX DENSITY NEAR MAGNETS AND COILS

- For fields due to magnets or steady fields due to direct current in coils, use a **Hall probe**. This can be calibrated by placing it in a field of known magnetic flux density.

- For varying magnetic fields produced by alternating currents in coils, use a **search coil** (a coil with small radius but hundreds of turns). The maximum **induced voltage** is proportional to the maximum magnetic flux density. Measure induced voltage using an oscilloscope.

**MUST REMEMBER**

The large face of the Hall probe and the plane of the search coil must be at right angles to the field being measured.

# FORCE ON A MOVING CHARGED PARTICLE

Direction of the force for electrons (negative charge) moving as shown

Force on a wire carrying a current is the resultant of the forces on individual charged particles (electrons) moving along the wire.

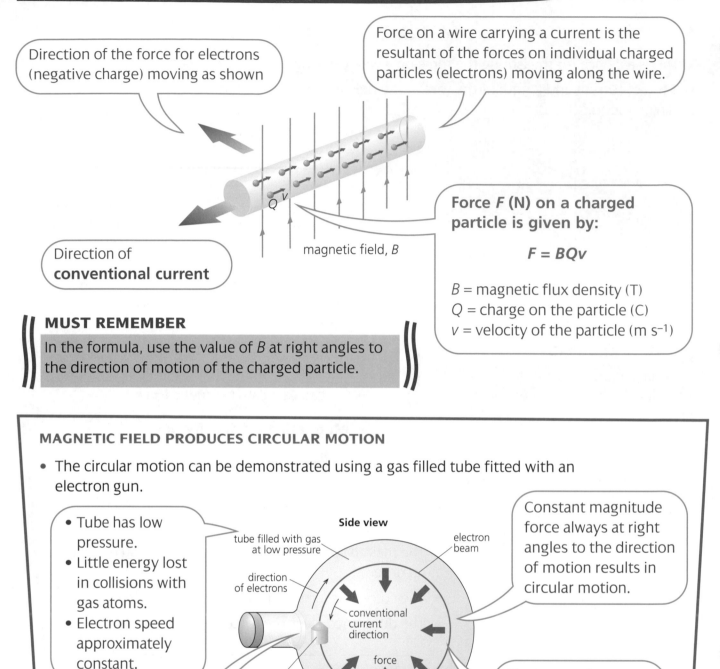

Direction of **conventional current**

magnetic field, $B$

**Force $F$ (N) on a charged particle is given by:**

$$F = BQv$$

$B$ = magnetic flux density (T)
$Q$ = charge on the particle (C)
$v$ = velocity of the particle (m s$^{-1}$)

**MUST REMEMBER**

In the formula, use the value of $B$ at right angles to the direction of motion of the charged particle.

---

**MAGNETIC FIELD PRODUCES CIRCULAR MOTION**

- The circular motion can be demonstrated using a gas filled tube fitted with an electron gun.

- Tube has low pressure.
- Little energy lost in collisions with gas atoms.
- Electron speed approximately constant.

**Side view**

tube filled with gas at low pressure

direction of electrons

conventional current direction

electron beam

force

electron gun

Magnetic field into plane of paper

Constant magnitude force always at right angles to the direction of motion results in circular motion.

Electron gun produces electrons of velocity $v$.

Electron path – gas atoms emit light following excitation by colliding electrons.

- **Force on charged particle moving at right angles to a magnetic field is equal to the centripetal force.**

$$BQv = \frac{mv^2}{r} \text{ so } r = \frac{mv}{BQ}$$

$Q$ = charge of particle ($-1.6 \times 10^{-19}$ C for an electron)
$m$ = mass of particle ($9.1 \times 10^{-31}$ kg for an electron)

- Electrons exhibit particle behaviour when influenced by electric and magnetic fields.
  - Classical mechanics can be used to explain the motion.
  - Only particles have mass and can carry charge.

# SPECIFIC CHARGE OF AN ELECTRON

- **Specific charge** of an electron is the ratio $\dfrac{\text{electron charge}}{\text{electron mass}}$
- Use **Millikan's experiment** to find the electron charge.
- Use value of $\dfrac{e}{m}$ to find the electron mass.

---

**EXPERIMENT TO MEASURE $\dfrac{e}{m}$**

- Using the gas filled tube on page 80, measure:
  – accelerating voltage in electron gun
  – radius of the circular path (from diameter)
  – magnetic flux density.

> The electron velocity $v$ can also be found using a method that balances the forces due to electric and magnetic fields:
> $$v = \frac{E}{B} \quad \text{(see 'Using fields', page 83)}$$
> $$\frac{e}{m} = \frac{E}{B^2 r}$$

- The equation for the electron gun is $\frac{1}{2}mv^2 = eV$
  so electron velocity $v = \sqrt{\dfrac{2eV}{m}}$

- For the circular path of the electron
  $$Bev = \frac{mv^2}{r} \qquad \text{so} \quad \frac{e}{m} = \frac{v}{Br}$$
  $$\frac{e}{m} = \frac{1}{Br}\sqrt{\frac{2eV}{m}} \quad \text{so} \quad \left(\frac{e}{m}\right)^2 = \frac{2V}{B^2 r^2}\left(\frac{e}{m}\right)$$
  $$\frac{e}{m} = \frac{2V}{B^2 r^2}$$

> For a stationary or slow-moving electron the **specific charge**:
> $$\frac{e}{m} = 1.8 \times 10^{11}\,\text{C kg}^{-1}$$
> (For very fast moving electrons there is a relativistic increase in mass, so $\dfrac{e}{m}$ decreases.)

**SPECIFIC CHARGE OF HYDROGEN IONS**

- The specific charge of a hydrogen ion is $9.4 \times 10^6\,\text{C kg}^{-1}$
- This is 1/1800 of the specific charge of an electron.
- The charge on a hydrogen ion is $+e$ so the mass must be 1800 times greater.
- A hydrogen ion is a proton so proton mass $\approx 1800 \times$ electron mass

---

**WORKED EXAMPLES**

**1** An electron is accelerated from rest by a p.d. of 3.5 kV.
Calculate:
(a) the velocity of the electron
(b) the flux density of the field that would cause the electron to move in a circular path of radius 0.050 m.
For an electron $\dfrac{e}{m} = 1.8 \times 10^{11}\,\text{C kg}^{-1}$

1 (a) $\frac{1}{2}mv^2 = eV$
$$v = \sqrt{\frac{2eV}{m}} = \sqrt{2 \times 1.8 \times 10^{11} \times 3500} = 3.5 \times 10^7\,\text{m s}^{-1}$$

(b) $Bev = \dfrac{mv^2}{r}$ so $B = \dfrac{mv}{er}$
$$B = \frac{3.5 \times 10^7}{1.8 \times 10^{11} \times 0.050} = 3.9\,\text{mT}$$

**2** The electrons in the gas filled tube lose energy to produce the visible path. State and explain the effect this has on the path of the electron.

> This also explains the **spiral path** seen when electrons and positrons are detected by a cloud or bubble chamber.

2 The radius of the path will become smaller as the electron travels in the tube producing a spiral path. The kinetic energy and hence the velocity of the electron fall. The radius of the path is given by $r = \dfrac{mv}{Be}$. So as $v$ decreases, the radius must also decrease since $m$, $B$ and $e$ are constant.

# USING FIELDS

## D.C. MOTOR

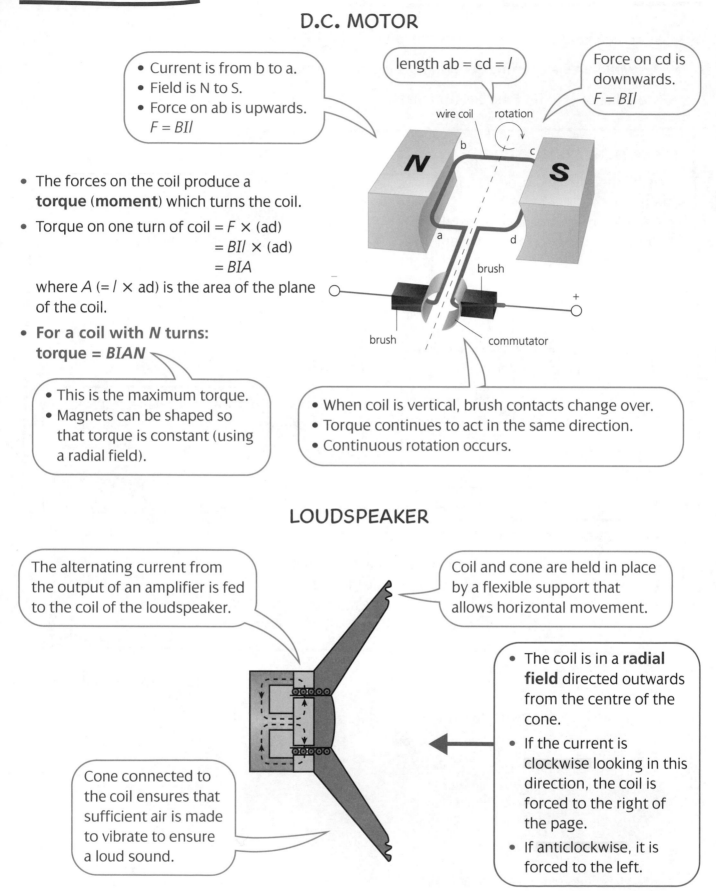

- Current is from b to a.
- Field is N to S.
- Force on ab is upwards.
  $F = BIl$

length ab = cd = $l$

Force on cd is downwards.
$F = BIl$

- The forces on the coil produce a **torque** (**moment**) which turns the coil.
- Torque on one turn of coil = $F \times$ (ad)
  $\qquad\qquad\qquad\quad = BIl \times$ (ad)
  $\qquad\qquad\qquad\quad = BIA$

  where $A$ (= $l \times$ ad) is the area of the plane of the coil.
- **For a coil with $N$ turns:**
  **torque = $BIAN$**

- This is the maximum torque.
- Magnets can be shaped so that torque is constant (using a radial field).

- When coil is vertical, brush contacts change over.
- Torque continues to act in the same direction.
- Continuous rotation occurs.

## LOUDSPEAKER

The alternating current from the output of an amplifier is fed to the coil of the loudspeaker.

Coil and cone are held in place by a flexible support that allows horizontal movement.

- The coil is in a **radial field** directed outwards from the centre of the cone.
- If the current is clockwise looking in this direction, the coil is forced to the right of the page.
- If anticlockwise, it is forced to the left.

Cone connected to the coil ensures that sufficient air is made to vibrate to ensure a loud sound.

- Alternating currents produce **forced vibrations** of the coil and cone.
- The sound has the same frequency as the alternating current.
- The output of the loudspeaker depends on the frequency as well as the magnitude of the current in the coil. It produces maximum sound output at its **resonant frequency**

# MASS SPECTROMETER

- A **mass spectrometer** (**spectrograph**) is used to find the masses of ions.
- An ion is an atom that has lost one or more electrons, so it carries a positive charge.
- Since electrons have low mass, the mass of an ion ≈ mass of an atom ≈ mass of the nucleus.

Sample of the element is evaporated in an oven.

The atoms move into a strong electric field, which produces positive ions.

The ions are accelerated and collimated into a fine beam.

All ions have the same velocity on leaving the velocity selector.

A magnetic field (out of the paper) deflects the ions – the magnetic force provides the centripetal force.

Only ions with a particular mass are deflected into the detector.
(Heavy ions are not deflected enough and light ions are deflected too much.)

## THE VELOCITY SELECTOR

electric field $E_S$   + magnetic field into plane $B_S$

$F_B = B_S Q v$
$F_E = E_S Q$

- When $F_B = F_E$ there is no resultant force on the ion, so it travels straight on.
- Since $B_S Q v = E_S Q$ this happens when:

$$v = \frac{E_S}{B_S}$$

($E_S$ and $B_S$ can be calculated or measured.)

- In the deflecting region:

$$BQv = \frac{mv^2}{r} \quad \text{so} \quad m = \frac{BrQ}{v}$$

- For this arrangement, $v$ and $r$ are constant. $Q$ is usually $+e$ or $+2e$.
- Provided that all ions have the same charge, **mass ∝ flux density**

- In this type of mass spectrometer, ions are deflected onto a photographic plate.
- **Heavier atoms** move in a path of larger radius.
- In the ion separator, $B$ and $v$ are constant so if $Q$ is the same for all ions then mass ∝ radius.

(Radius of path is **halved** for ions of the **same mass and velocity** but **twice the charge**.)

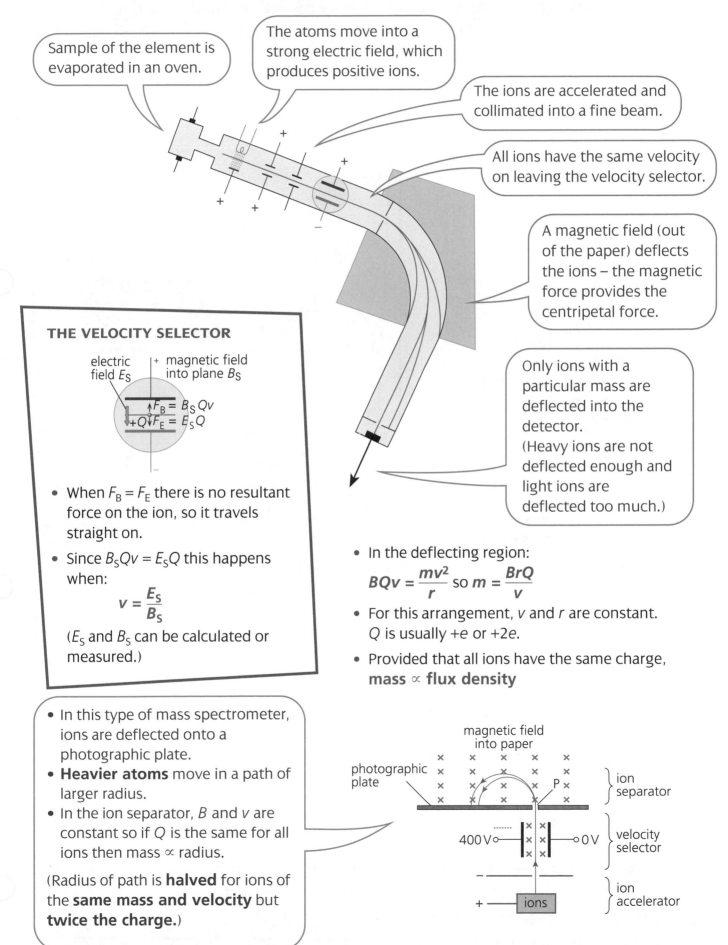

# PARTICLE ACCELERATORS

- Charged particles are accelerated to high speed by electric fields, for example:
  – using an electron gun in a television tube or cathode ray oscilloscope
  – using a high voltage generator such as a van de Graaff generator.
- To produce very high energy particles, a charged particle has to be accelerated many times.
- Accelerators are used for research, for medical treatment and for production of radioisotopes for industrial and medical use.

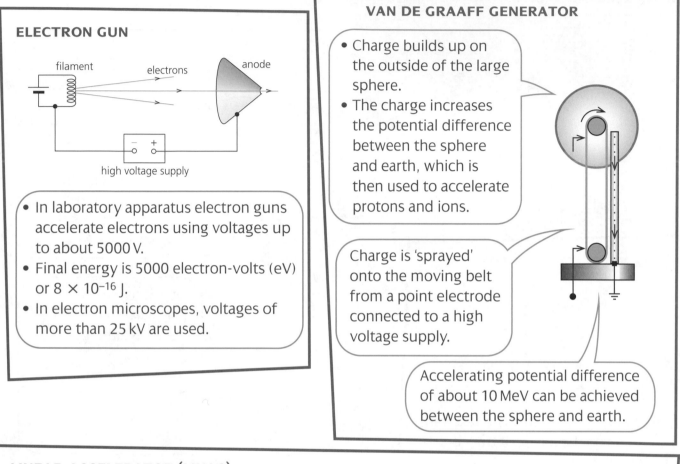

## ELECTRON GUN

filament
electrons
anode

high voltage supply

- In laboratory apparatus electron guns accelerate electrons using voltages up to about 5000 V.
- Final energy is 5000 electron-volts (eV) or $8 \times 10^{-16}$ J.
- In electron microscopes, voltages of more than 25 kV are used.

## VAN DE GRAAFF GENERATOR

- Charge builds up on the outside of the large sphere.
- The charge increases the potential difference between the sphere and earth, which is then used to accelerate protons and ions.

Charge is 'sprayed' onto the moving belt from a point electrode connected to a high voltage supply.

Accelerating potential difference of about 10 MeV can be achieved between the sphere and earth.

## LINEAR ACCELERATOR (LINAC)

- Particles are accelerated in stages and have final energies of up to 10 GeV ($10 \times 10^9$ eV).

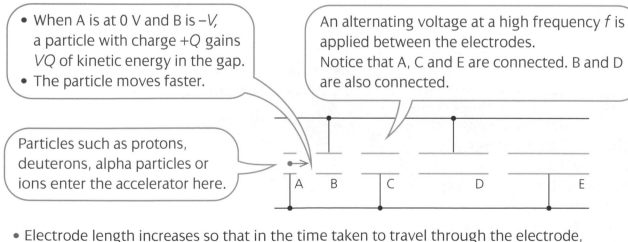

- When A is at 0 V and B is $-V$, a particle with charge $+Q$ gains $VQ$ of kinetic energy in the gap.
- The particle moves faster.

An alternating voltage at a high frequency $f$ is applied between the electrodes.
Notice that A, C and E are connected. B and D are also connected.

Particles such as protons, deuterons, alpha particles or ions enter the accelerator here.

A   B   C   D   E

- Electrode length increases so that in the time taken to travel through the electrode, the polarity changes.
- **Length of electrode = Velocity of particle entering electrode** $\times \dfrac{1}{2f}$
- Particle meets an accelerating field at the next gap so it is accelerated again.
- **Final energy = number of gaps $\times VQ$**

# CYCLOTRON

- A **cyclotron** uses a magnetic field to produce circular motion resulting in a more compact high energy accelerator.

- A cyclotron is used to accelerate heavy particles like protons and ions to energies of 20 MeV.

Charged particles are injected near the centre of the cyclotron. As velocity increases, the path spirals outwards.

An alternating voltage is connected across the two Dee-shaped electrodes.
The frequency is chosen so that the particle always travels in an accelerating field when it crosses the gap between the Dees.

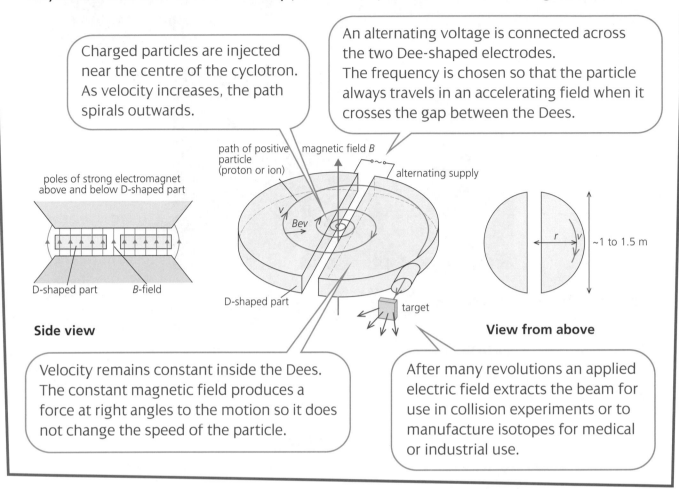

poles of strong electromagnet above and below D-shaped part

D-shaped part    B-field

**Side view**

path of positive particle (proton or ion)    magnetic field B

alternating supply

Bev

D-shaped part    target

r    v    ~1 to 1.5 m

**View from above**

Velocity remains constant inside the Dees. The constant magnetic field produces a force at right angles to the motion so it does not change the speed of the particle.

After many revolutions an applied electric field extracts the beam for use in collision experiments or to manufacture isotopes for medical or industrial use.

## WHAT FIXES THE CYCLOTRON FREQUENCY?

- For acceleration at each gap, the polarity of the Dees must change during the time taken for the particle to travel half a revolution.

- This occurs when the frequency of revolution matches the frequency of the alternating p.d.

- This frequency is the **cyclotron frequency**.

- For speed $v$, time $t$ to travel half a revolution $= \dfrac{\pi r}{v}$

  For a particle carrying a charge $e$, when radius of path is $r$,

  $Bev = \dfrac{mv^2}{r}$ so $\dfrac{r}{v} = \dfrac{m}{Be}$ which is constant.

  $t = \dfrac{\pi m}{Be}$

  Period $T$ of revolution $= 2t$ and $f = \dfrac{1}{T}$ so

  $$\text{cyclotron frequency, } f = \frac{Be}{2\pi m}$$

- The frequency is fixed by the flux density of the field and the specific charge of the particle being accelerated.

- The final energy is restricted because of the relativistic increase in the mass. This causes the synchronisation of the alternating voltage and the frequency of the circular motion to break down.

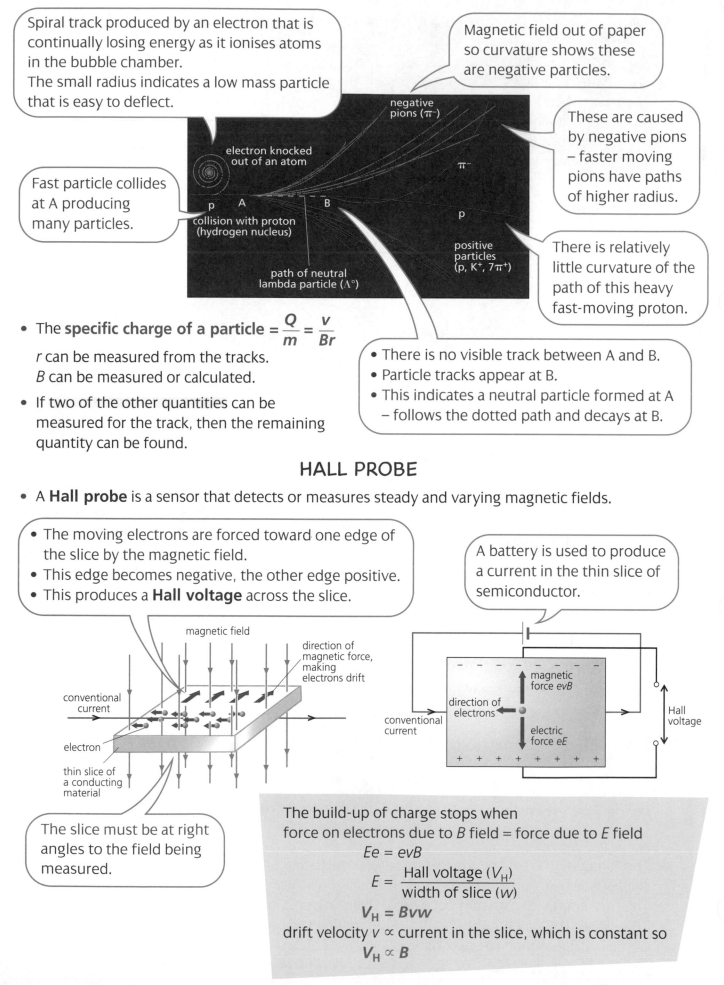

# ANALYSING PARTICLE TRACKS

- Accelerated particles colliding with matter in a cloud or bubble chamber produce other particles.
- Charged particles produce curved tracks when a magnetic field is applied to the chamber.

Spiral track produced by an electron that is continually losing energy as it ionises atoms in the bubble chamber.
The small radius indicates a low mass particle that is easy to deflect.

Magnetic field out of paper so curvature shows these are negative particles.

These are caused by negative pions – faster moving pions have paths of higher radius.

Fast particle collides at A producing many particles.

There is relatively little curvature of the path of this heavy fast-moving proton.

- The **specific charge of a particle** $= \dfrac{Q}{m} = \dfrac{v}{Br}$

  $r$ can be measured from the tracks.
  $B$ can be measured or calculated.

- If two of the other quantities can be measured for the track, then the remaining quantity can be found.

- There is no visible track between A and B.
- Particle tracks appear at B.
- This indicates a neutral particle formed at A – follows the dotted path and decays at B.

# HALL PROBE

- A **Hall probe** is a sensor that detects or measures steady and varying magnetic fields.

- The moving electrons are forced toward one edge of the slice by the magnetic field.
- This edge becomes negative, the other edge positive.
- This produces a **Hall voltage** across the slice.

A battery is used to produce a current in the thin slice of semiconductor.

The slice must be at right angles to the field being measured.

The build-up of charge stops when
force on electrons due to $B$ field = force due to $E$ field
$$Ee = evB$$
$$E = \frac{\text{Hall voltage } (V_H)}{\text{width of slice } (w)}$$
$$V_H = Bvw$$
drift velocity $v \propto$ current in the slice, which is constant so
$$V_H \propto B$$

# COMBINING FIELDS

- **Magnetic flux density *B*** is a **vector** quantity.
- When the fields are in the same direction, the flux densities are added, and when in opposite directions they are subtracted.

## WHAT IS A NEUTRAL POINT?

- At a **neutral point**, there is no resultant force due to magnetic fields.
- There is no magnetic flux density at a neutral point.

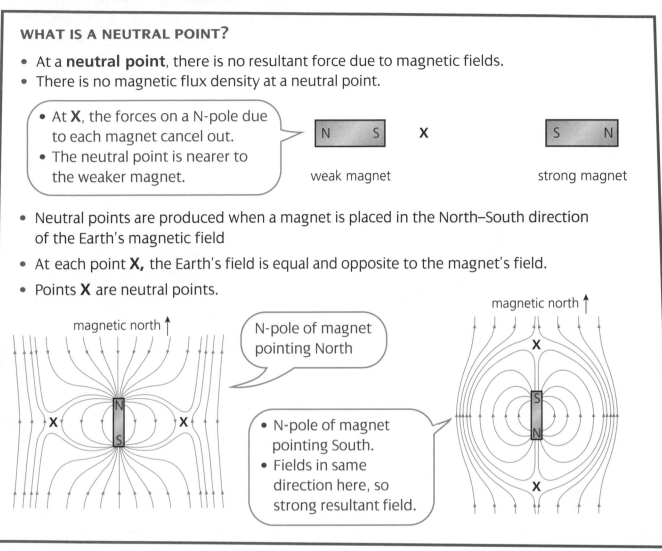

- At **X**, the forces on a N-pole due to each magnet cancel out.
- The neutral point is nearer to the weaker magnet.

weak magnet

strong magnet

- Neutral points are produced when a magnet is placed in the North–South direction of the Earth's magnetic field
- At each point **X,** the Earth's field is equal and opposite to the magnet's field.
- Points **X** are neutral points.

N-pole of magnet pointing North

- N-pole of magnet pointing South.
- Fields in same direction here, so strong resultant field.

## INTERACTING FIELDS DUE TO CURRENTS IN WIRES

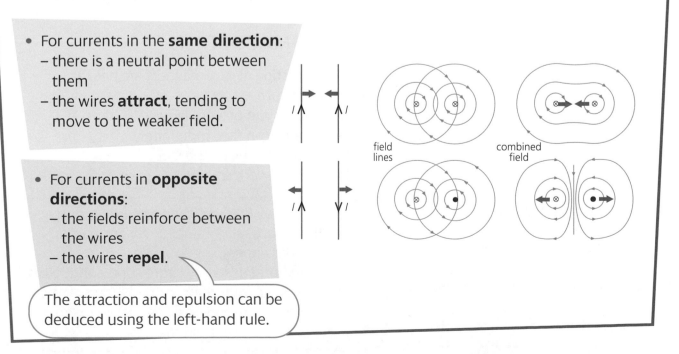

- For currents in the **same direction**:
  – there is a neutral point between them
  – the wires **attract**, tending to move to the weaker field.

- For currents in **opposite directions**:
  – the fields reinforce between the wires
  – the wires **repel**.

field lines

combined field

The attraction and repulsion can be deduced using the left-hand rule.

# MAGNETIC MATERIALS AND MAGNETS

- Some atoms produce a resultant magnetic field due to motion of electrons that orbit the nucleus – they are like tiny magnets.

- In ferromagnetic materials (iron, nickel and cobalt), the atoms are aligned in small **domains** but normally a lump of the material has little resultant magnetism.

- The domains can be aligned by placing the material in a strong magnetic field produced by a **solenoid**.

The **ease of reversing magnetism** with little energy loss makes the **soft** material suitable for electromagnets and transformer cores.

- **Hysteresis** curve for sample (initially unmagnetised) taken through cycle ABCDEFGB.
- The area enclosed is proportional to the energy that becomes internal energy of the sample during the cycle.

**Magnetic saturation** All domains aligned →

Domains random

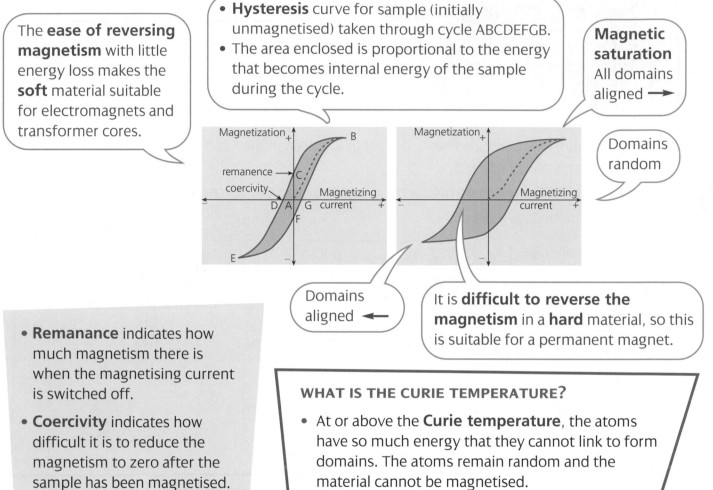

Domains aligned ←

It is **difficult to reverse the magnetism** in a **hard** material, so this is suitable for a permanent magnet.

- **Remanance** indicates how much magnetism there is when the magnetising current is switched off.

- **Coercivity** indicates how difficult it is to reduce the magnetism to zero after the sample has been magnetised.

## WHAT IS THE CURIE TEMPERATURE?

- At or above the **Curie temperature**, the atoms have so much energy that they cannot link to form domains. The atoms remain random and the material cannot be magnetised.

## SUPERCONDUCTOR MAGNETS

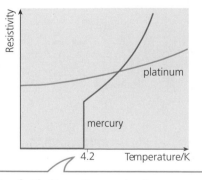

- **Resistivity** of metals decreases as temperature falls.
- At the **transition temperature** the resistivity of a **superconductor** becomes **zero**.

- Superconducting electromagnets are held at low temperature.
  - The currents can be very large.
  - The currents can flow indefinitely.
  - There is no heating of the wire.
  - The fields are very steady.

- The high steady fields of superconductor magnets make them particularly suitable for applications such as MRI scanners.
- The search for high temperature superconductors is important for applications such as transmission of electrical energy and possible uses in magnetic levitation in transport systems.

# ELECTROMAGNETIC INDUCTION

## LAWS OF ELECTROMAGNETIC INDUCTION

### FARADAY'S LAWS

- When a conductor cuts through magnetic flux or when the flux linking a circuit changes, an e.m.f. is induced.
- The magnitude of the induced e.m.f. is proportional to the rate at which flux is cut or the rate of change of flux linkage.

### LENZ'S LAW

- The direction of the current produced by the induced e.m.f. opposes the change that is producing it.

### WHY IS THERE AN INDUCED E.M.F. WHEN A WIRE CUTS FLUX?

- Metal wire contains positive ions (fixed in lattice) and free electrons.
- As wire moves, the charged ions and electrons experience a force in opposite directions.
- The ions cannot move but electrons move toward end X (see 'Force on a moving charged particle', page 80).
- End X becomes negatively charged and end Y positively charged.
- The potential difference between ends is the induced e.m.f.

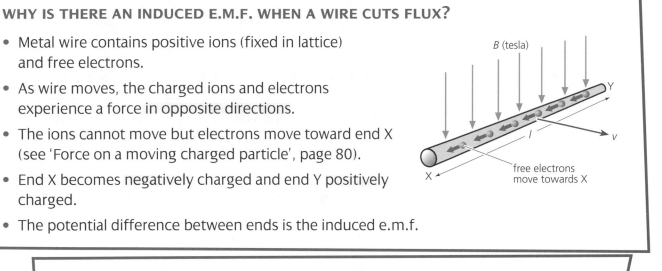

### HOW BIG IS THE INDUCED E.M.F. WHEN A WIRE CUTS FLUX?

- The e.m.f. $V$ is induced between the ends X and Y of the wire.
- The wire has a length $l$ and moves at a velocity $v$ at right angles to the flux.
- Electric field strength $E$ in the wire due to the induced e.m.f. $= \dfrac{V}{l}$
- Electrons stop moving toward the end X when:
  Force on an electron due to the $E$ field in the wire = Force due to the $B$ field

$$eE = Bev \quad \text{so} \quad \frac{Ve}{l} = BeV$$

**Induced e.m.f. $V = Blv$**

- This e.m.f. is induced whether or not there is a complete circuit.
- If there is a complete circuit, charge continues to flow and there is an induced current.

- This voltage cannot be measured using a voltmeter in the aircraft.
- The wires connecting the voltmeter to the wing-tips would have the same e.m.f. induced in them.
- The two e.m.f.s would be opposite to each other in the circuit and therefore cancel.

**WORKED EXAMPLE**

The wing-span of an aircraft is 55 m. Calculate the induced e.m.f. between the wing-tips of the aircraft when it is moving at 500 km h⁻¹ at a position where the vertical component of the Earth's magnetic field is $1.5 \times 10^{-4}$ T.

$$\text{Speed of aircraft} = \frac{500 \times 10^3}{60 \times 60} = 139 \, \text{m s}^{-1}$$

$$\begin{aligned}
\text{Induced e.m.f.} &= Blv \\
&= 1.5 \times 10^{-4} \times 55 \times 139 \, \text{V} \\
&= 1.15 \, \text{V}
\end{aligned}$$

# MAGNETIC FLUX $\phi$

- Magnetic flux is an abstract concept that helps describe the behaviour of magnetic fields.
- The magnetic field is represented by **flux lines** – the greater the concentration of lines, the stronger the field.

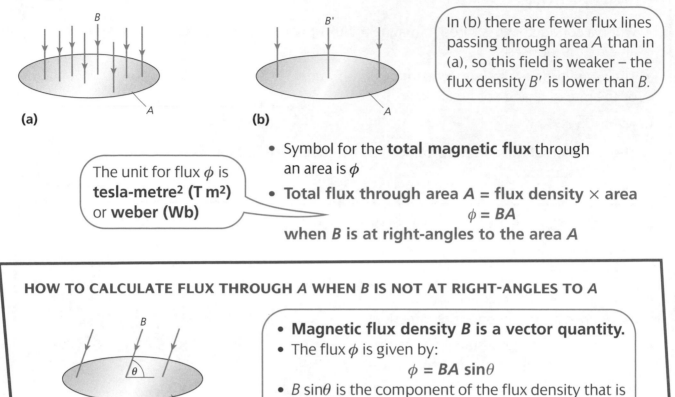

**(a)**

**(b)**

In (b) there are fewer flux lines passing through area $A$ than in (a), so this field is weaker – the flux density $B'$ is lower than $B$.

The unit for flux $\phi$ is **tesla-metre² (T m²)** or **weber (Wb)**

- Symbol for the **total magnetic flux** through an area is $\phi$
- Total flux through area $A$ = flux density × area
$$\phi = BA$$
when $B$ is at right-angles to the area $A$

---

**HOW TO CALCULATE FLUX THROUGH $A$ WHEN $B$ IS NOT AT RIGHT-ANGLES TO $A$**

- **Magnetic flux density $B$ is a vector quantity.**
- The flux $\phi$ is given by:
$$\phi = BA \sin\theta$$
- $B \sin\theta$ is the component of the flux density that is at right-angles to $A$.

**MUST REMEMBER**

$\theta$ is the angle between the direction of the flux and the plane of area $A$.

# MAGNETIC FLUX LINKAGE $\Phi$

- The magnetic flux of a magnetic field may pass through a circuit.
- **The magnetic flux linkage is the total magnetic flux that is linked to the circuit. Flux linkage $\Phi = \phi \times$ total area through which flux passes**
- The total area is increased when the circuit is in the form of a coil.

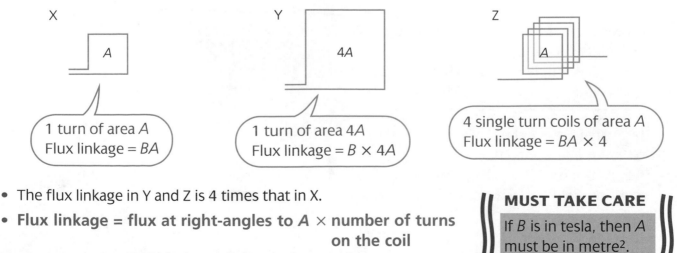

X

$A$

1 turn of area $A$
Flux linkage = $BA$

Y

$4A$

1 turn of area $4A$
Flux linkage = $B \times 4A$

Z

$A$

4 single turn coils of area $A$
Flux linkage = $BA \times 4$

- The flux linkage in Y and Z is 4 times that in X.
- **Flux linkage = flux at right-angles to $A$ × number of turns on the coil**
$$\Phi = BAN \text{ ($B$ is at right-angles to $A$)}$$

**MUST TAKE CARE**

If $B$ is in tesla, then $A$ must be in metre².

# DEMONSTRATING ELECTROMAGNETIC INDUCTION

## FLUX CUTTING EXPERIMENT

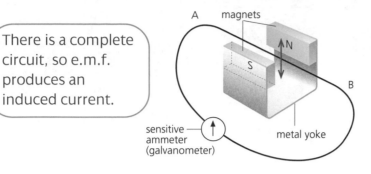

There is a complete circuit, so e.m.f. produces an induced current.

sensitive ammeter (galvanometer)

metal yoke

**Demonstrating Faraday's 1st law**
E.m.f. induced only when wire moves through the field

**Demonstrating Faraday's 2nd law**
When the wire is moved faster, the reading is higher so there is a higher induced e.m.f.

### HOW DOES LENZ'S LAW APPLY?

- When wire is moving downwards, the meter deflects indicating an induced e.m.f.
- The induced current is from B to A in the wire that is in the field.
- This produces an upward force on the wire (shown using left-hand rule).
- The downward motion produces the current – the force acts upwards to try to oppose this downward motion – as Lenz's law states.

- Because the force opposes the motion, work has to be done to move the wire in the field. This work is done by a source of energy.
- For a wire falling freely, the force to produce the motion is the gravitational force and some of the change in gravitational potential energy is transferred to electrical energy.
- **Lenz's law is a consequence of the principle of conservation of energy.**

## INDUCING E.M.F.S IN A COIL USING A MOVING MAGNET

**Demonstrating Faraday's 1st law**
When the magnet is moved towards the coil, the flux linkage increases.

**Demonstrating Faraday's 2nd law**
The faster the movement of the magnet, the greater the **rate of change of flux linkage**, so a higher induced e.m.f.

This end behaves like a N pole

Now it behaves like a S pole

**DEMONSTRATING LENZ'S LAW**

As the magnet moves toward the coil, the induced current makes the nearer end of the coil a N-pole, which repels the magnet so opposing the motion.

As the magnet moves away from the coil, the induced current makes the nearer end of the coil a S-pole, which attracts the magnet so opposing the motion.

# CHANGING FLUX WITHOUT MOTION

When the switch is opened or when it is closed, an e.m.f. is induced in this coil which produces an induced current.

The flux in this coil increases.

When the switch is closed, this coil creates magnetic flux.

The flux decreases again when the switch is opened.

## DEMONSTRATING LENZ'S LAW

- The direction of the induced current is in opposite directions when the current is switched on and off.

- The induced current is in the opposite direction to that in the right-hand coil when switching on.
- An increasing flux creates the induced current. The induced current creates flux in the opposite direction to try to prevent the increase.

- The induced current is in the same direction as that in the right-hand coil when switching off.
- A decreasing flux creates the induced current. The induced current creates flux in the same direction to try to reduce the decrease.

**MUST REMEMBER**

E.m.f. is produced only when the **flux is changing**, i.e. when the current in the right-hand coil is increasing or decreasing.

# DETECTING RADIO WAVES

electric field (E-field)

magnetic field (B-field)

**End-on view of an electromagnetic wave**

E-field

B-field

This type of aerial detects the voltage variation produced by the varying E-field.

- A coil orientated this way has currents induced in it by the varying magnetic field.
- A ferrite core increases the induced current.

- The wave from a transmitter is an electromagnetic wave.
- An e-m wave consists of interlinked electric and magnetic fields which oscillate at right angles to each other.

## SPEED OF E-M RADIATION

- Maxwell showed that the **speed** depends on the **permittivity** and **permeability** of the medium through which the wave travels.

- **Speed in free space** $c = \sqrt{\dfrac{1}{\mu_0 \varepsilon_0}}$

- Speed of radio waves in coaxial cable is lower. The wave travels in the polythene insulator. Permeability is the same as free space but the permittivity of polythene is higher by a factor of 2.2.

- Speed along coaxial cable is $\dfrac{1}{\sqrt{2.2}} c = 0.67c$

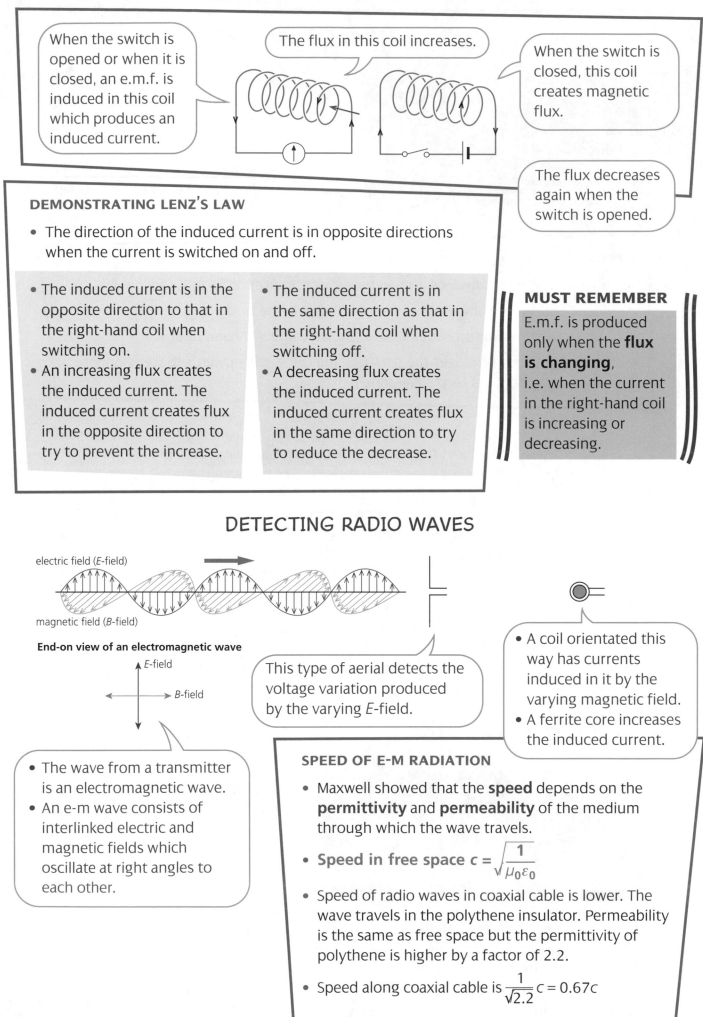

# MAGNITUDE OF INDUCED E.M.F.S

- From Faraday's law, induced e.m.f. is proportional to the rate of change of flux linkage, $\left(\dfrac{\Delta\Phi}{\Delta t}\right)$

- Flux linkage $\Phi$ through area $A = BA$ so **magnitude of induced e.m.f.** $= \dfrac{N\Delta(BA)}{\Delta t}$

**In an alternator**
$B$ is constant and the area through which the flux passes changes as the coil rotates.

**In a transformer**
$A$ is constant and the magnitude and direction of the flux change as an alternating current changes.

**DEFINING THE UNIT OF FLUX**

- The unit of $\Phi$ is the weber (Wb).

- When the flux linkage changes at a rate of 1 weber per second, the induced e.m.f. is 1 V.

- Since $\Phi = BA$, 1 Wb = 1 T m$^2$

**WORKED EXAMPLES**

**1** A coil has 350 turns and an area of 4.5 cm$^2$. When the magnetic flux though it changes, an e.m.f. of 0.13 V is recorded between the terminals.
Calculate the rate of change of flux density that produces this e.m.f.

Area of coil $= 4.5 \times 10^{-4}$ m$^2$

Induced e.m.f. $= \dfrac{NA\Delta(B)}{\Delta t}$

$0.13 = 350 \times 4.5 \times 10^{-4} \times \dfrac{\Delta B}{\Delta t}$

$\dfrac{\Delta B}{\Delta t} = 0.83$ T s$^{-1}$

**2** The graph shows how the flux density through an 1100 turn coil of area 8.0 cm$^2$ varies with time.

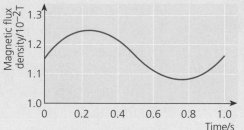

(a) Calculate the magnitude of maximum induced e.m.f. between the terminals.
(b) Sketch a graph showing how the induced e.m.f. varies with time.

(a) The gradient at any time is the rate of change of flux density. This is proportional to the induced e.m.f. Maximum rate of change of flux density is at 0.50 s, found by drawing tangent.

$\left(\dfrac{\Delta B}{\Delta t}\right)_{max} \approx 0.0055$ T s$^{-1}$

Maximum induced e.m.f. $= 0.0055 \times 1100 \times 8.0 \times 10^{-4}$
$\qquad\qquad = 4.8$ mV

(b)

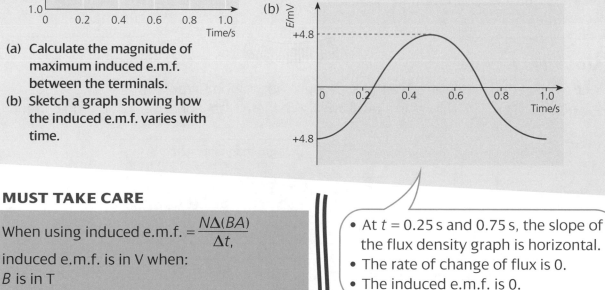

**MUST TAKE CARE**

When using induced e.m.f. $= \dfrac{N\Delta(BA)}{\Delta t,}$

induced e.m.f. is in V when:
$B$ is in T
$A$ is in m$^2$
$t$ is in s
$N$ is the **total number of turns** on the coil.

- At $t = 0.25$ s and 0.75 s, the slope of the flux density graph is horizontal.
- The rate of change of flux is 0.
- The induced e.m.f. is 0.

# EDDY CURRENTS

- **Eddy currents** are produced in a metal object such as a plate or block when the magnetic flux through it changes.
- There is no external circuit, but currents can circulate within the plate or metal block.

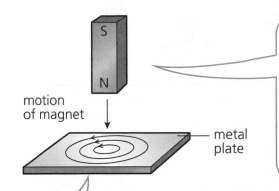

motion of magnet

metal plate

- When the N-pole of the magnet moves towards the plate, the circulating currents are anticlockwise.
- The currents produce a force that repels the magnet, opposing its motion (Lenz's law).
- When the magnet moves away, the currents are in the opposite direction producing a force that attracts the magnet.

- Currents in the plate **form complete loops**.
- The currents produce a rise in temperature of the plate due to $I^2R$ heating.

**MUST TAKE CARE**

Only use the term **eddy current** for a current that circulates in a metal plate or block.
The induced current in a coil caused by a changing magnetic field is not an eddy current.

## DAMPING MOTION USING EDDY CURRENTS

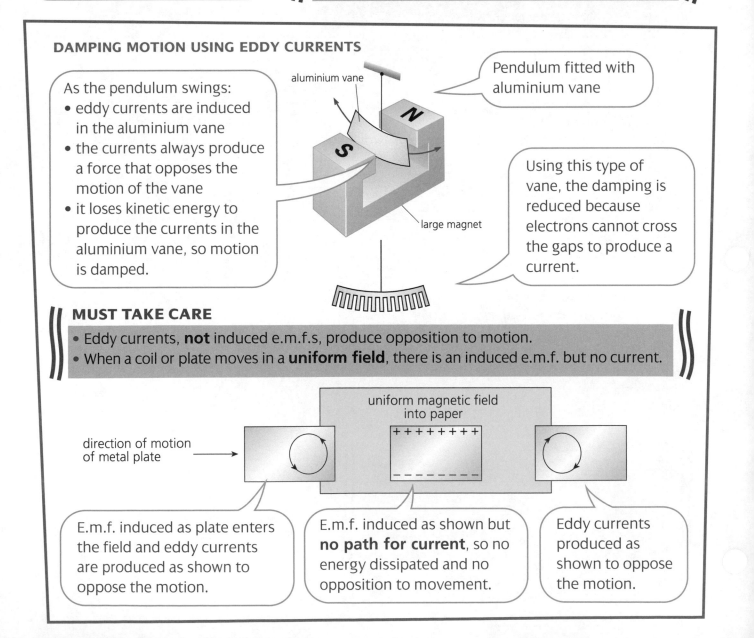

As the pendulum swings:
- eddy currents are induced in the aluminium vane
- the currents always produce a force that opposes the motion of the vane
- it loses kinetic energy to produce the currents in the aluminium vane, so motion is damped.

aluminium vane

Pendulum fitted with aluminium vane

large magnet

Using this type of vane, the damping is reduced because electrons cannot cross the gaps to produce a current.

**MUST TAKE CARE**

- Eddy currents, **not** induced e.m.f.s, produce opposition to motion.
- When a coil or plate moves in a **uniform field**, there is an induced e.m.f. but no current.

uniform magnetic field into paper

direction of motion of metal plate

+ + + + + + + +

– – – – – – – –

E.m.f. induced as plate enters the field and eddy currents are produced as shown to oppose the motion.

E.m.f. induced as shown but **no path for current**, so no energy dissipated and no opposition to movement.

Eddy currents produced as shown to oppose the motion.

# USING ELECTROMAGNETIC INDUCTION

## A.C. GENERATOR (ALTERNATOR)

- An a.c. generator consists of a coil rotating in a magnetic field.

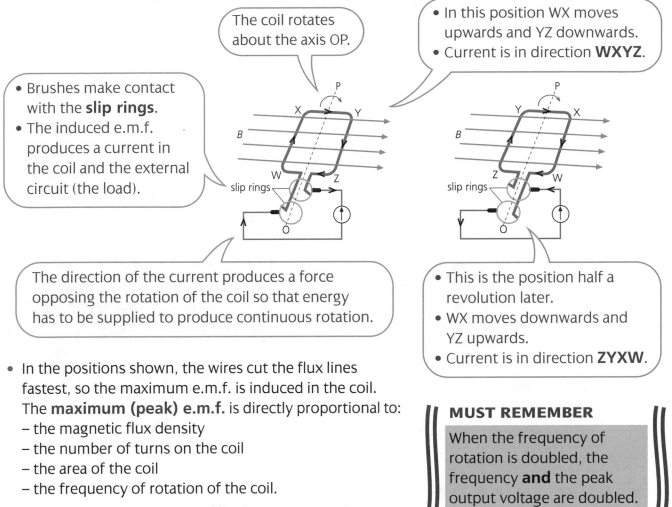

The coil rotates about the axis OP.

- In this position WX moves upwards and YZ downwards.
- Current is in direction **WXYZ**.

- Brushes make contact with the **slip rings**.
- The induced e.m.f. produces a current in the coil and the external circuit (the load).

The direction of the current produces a force opposing the rotation of the coil so that energy has to be supplied to produce continuous rotation.

- This is the position half a revolution later.
- WX moves downwards and YZ upwards.
- Current is in direction **ZYXW**.

- In the positions shown, the wires cut the flux lines fastest, so the maximum e.m.f. is induced in the coil. The **maximum (peak) e.m.f.** is directly proportional to:
  – the magnetic flux density
  – the number of turns on the coil
  – the area of the coil
  – the frequency of rotation of the coil.
- The frequency of the a.c. output is the same as the frequency of rotation of the coil.

**MUST REMEMBER**

When the frequency of rotation is doubled, the frequency **and** the peak output voltage are doubled.

---

### HOW DOES THE OUTPUT VOLTAGE VARY WITH TIME?

- As the coil rotates, the rate of change of flux linkage changes. This produces an output that varies sinusoidally with time.

The peak output is when the plane of the coil is parallel to the flux lines.

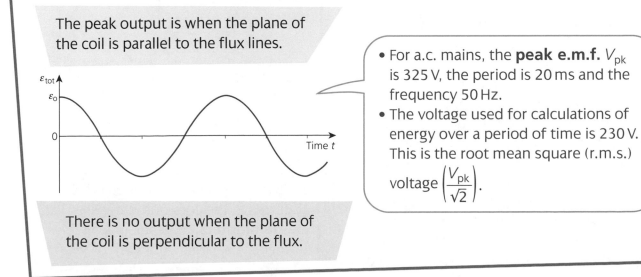

- For a.c. mains, the **peak e.m.f.** $V_{pk}$ is 325 V, the period is 20 ms and the frequency 50 Hz.
- The voltage used for calculations of energy over a period of time is 230 V. This is the root mean square (r.m.s.) voltage $\left(\dfrac{V_{pk}}{\sqrt{2}}\right)$.

There is no output when the plane of the coil is perpendicular to the flux.

# TRANSFORMERS

- When electrical energy is distributed to consumers, energy is lost in overcoming the electrical resistance of the cables.

- The losses can be reduced by reducing the current in the wires.

- By using transformers, it is possible to distribute the same power using a high voltage and low current so that losses are reduced.

## OPERATION OF A TRANSFORMER

The alternating input voltage $V_p$ to a **primary coil** produces a changing current $I_p$ in the primary coil.

The changing current in the primary coil produces a changing flux in the **soft iron core**.

- The changing flux in the core of the transformer produces an induced e.m.f. in the **secondary coil** of the transformer.
- This is the output voltage $V_s$
- When a load is connected, this produces a current $I_s$ in the secondary coil.

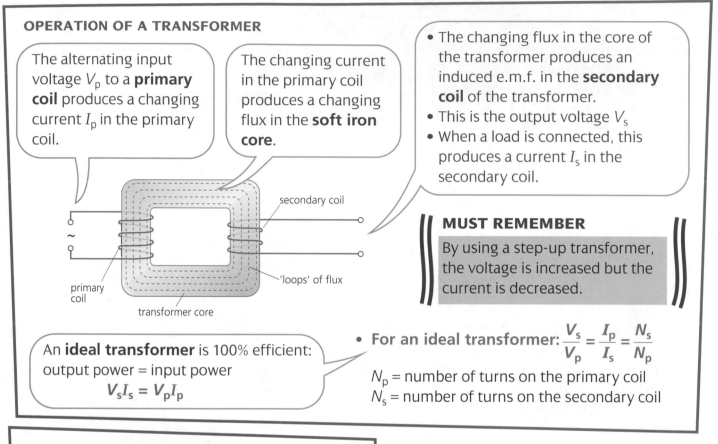

secondary coil

'loops' of flux

primary coil

transformer core

### MUST REMEMBER

By using a step-up transformer, the voltage is increased but the current is decreased.

An **ideal transformer** is 100% efficient:
output power = input power
$$V_s I_s = V_p I_p$$

- For an ideal transformer: $\dfrac{V_s}{V_p} = \dfrac{I_p}{I_s} = \dfrac{N_s}{N_p}$

$N_p$ = number of turns on the primary coil
$N_s$ = number of turns on the secondary coil

## EFFICIENCY OF PRACTICAL TRANSFORMERS

- In practice, some energy becomes internal energy of the transformer and is then transferred to the surroundings.

$$\text{Efficiency} = \frac{V_s I_s}{V_p I_p} \times 100\%$$

- Transformers are inefficient because of:
  - $I^2R$ **or copper loss** – current in primary and secondary coils produces heating due to resistance of wire.
  - **Flux loss** – flux produced by the primary current leaks from the core reducing the e.m.f. produced in the secondary.
  - **Eddy current loss** – eddy currents are produced in the core by the alternating current in the coils and result in heating due to electrical resistance of the core.
  - **Hysteresis loss** – the core is continually being magnetised one way and then the other resulting in a temperature rise in the core.

### WORKED EXAMPLE

An ideal transformer has 500 turns on the primary coil. The transformer produces a voltage of 6.0 V from a 230 V mains supply. The transformer operates a device with a power rating of 3.0 W. Calculate the number of turns on the secondary coil and the current drawn from the mains.

$\dfrac{V_s}{V_p} = \dfrac{N_s}{N_p}$ so $\dfrac{6.0}{230} = \dfrac{N_s}{500}$

Number of turns on the secondary = 13
Primary power = Secondary power
$$V_p I_p = 3.0\,\text{W}$$
Primary current = $\dfrac{3.0}{230} = 13\,\text{mA}$

To reduce the losses due to eddy currents, the core of the transformer is **laminated** (i.e. made of iron sheets insulated from each other).

# NUCLEAR STRUCTURE

A nucleus contains protons and neutrons – these are called **nucleons**.

nucleus ~10⁻¹⁴ m diameter

electron cloud ~10⁻¹⁰ m diameter

In a neutral atom, there are equal numbers of protons and electrons.

## HOW TO DESCRIBE A NUCLEUS

- A **nucleus** (or **nuclide**) is described by:
  - the number of protons in the nucleus – the **proton number Z**
  - the total number of protons and neutrons in the nucleus – the **nucleon number A**.

- The general symbol for a nucleus is

$$^A_Z X$$

X is the chemical symbol for the nucleus concerned.
**Number of neutrons = A – Z**

A hydrogen nucleus is a single proton. Its symbol is $^1_1 H$ or $^1_1 p$

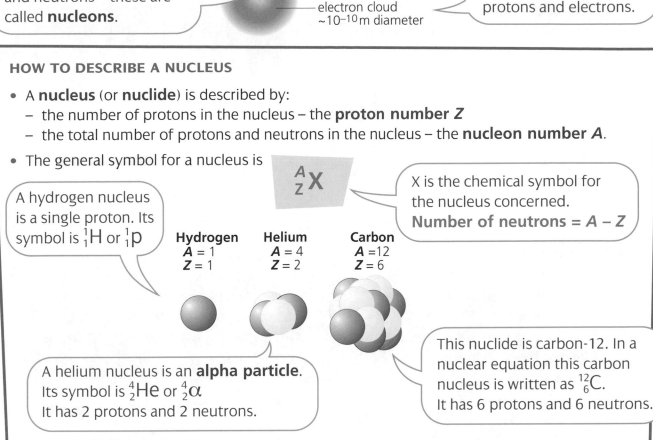

**Hydrogen**
A = 1
Z = 1

**Helium**
A = 4
Z = 2

**Carbon**
A = 12
Z = 6

A helium nucleus is an **alpha particle**. Its symbol is $^4_2 He$ or $^4_2 \alpha$. It has 2 protons and 2 neutrons.

This nuclide is carbon-12. In a nuclear equation this carbon nucleus is written as $^{12}_6 C$. It has 6 protons and 6 neutrons.

## WHAT ARE ISOTOPES?

- The proton number Z defines the name of the element and its chemical behaviour.
- **Isotopes** are atoms of the same element that have a different number of neutrons in their nuclei.

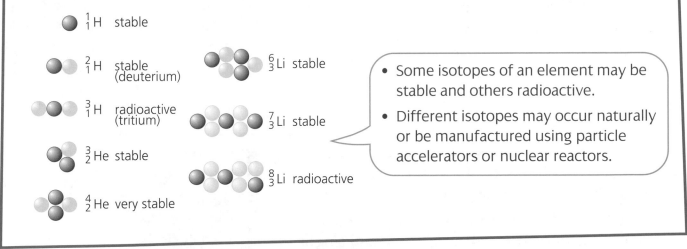

$^1_1 H$  stable

$^2_1 H$  stable (deuterium)

$^3_1 H$  radioactive (tritium)

$^3_2 He$ stable

$^4_2 He$ very stable

$^6_3 Li$  stable

$^7_3 Li$  stable

$^8_3 Li$  radioactive

- Some isotopes of an element may be stable and others radioactive.
- Different isotopes may occur naturally or be manufactured using particle accelerators or nuclear reactors.

## MASSES OF PARTICLES, NUCLEI AND ATOMS

- These are often given in terms of the atomic mass unit u.
  1 u is $\frac{1}{12}$ the mass of a carbon-12 nucleus = 1.660 540 2 × 10⁻²⁷ kg
- The mass of a proton is 1.007 276 5 u (= 1.672 623 × 10⁻²⁷ kg)
- The mass of a neutron is 1.008 664 0 u (= 1.674 927 × 10⁻²⁷ kg)

# DISCOVERY OF THE NUCLEUS

- This experiment was performed by Geiger and Marsden, working with Rutherford.

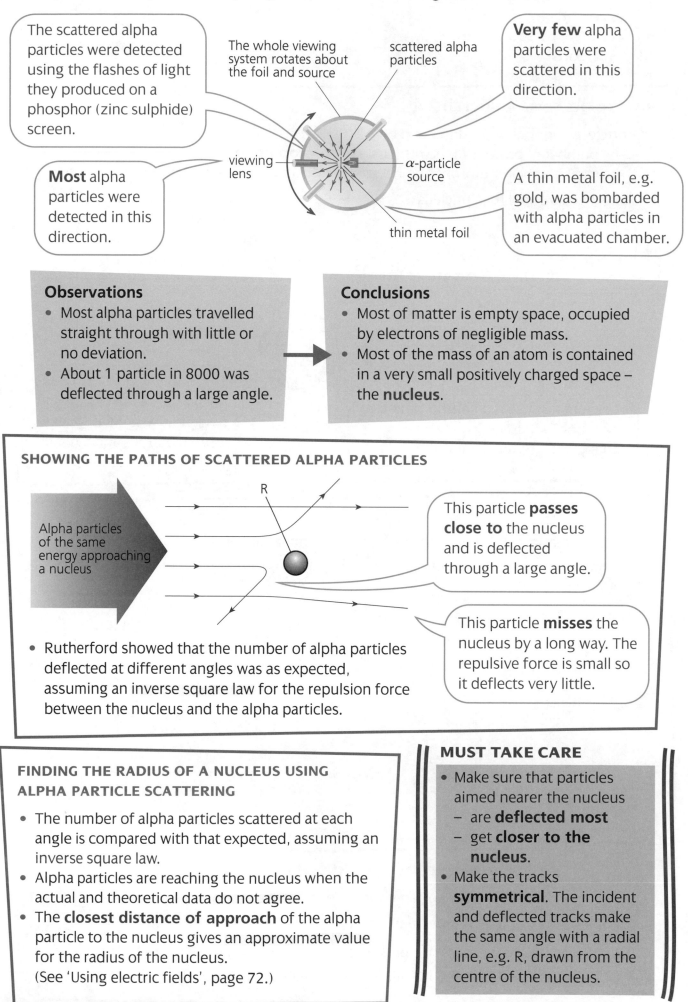

The scattered alpha particles were detected using the flashes of light they produced on a phosphor (zinc sulphide) screen.

The whole viewing system rotates about the foil and source

scattered alpha particles

**Very few** alpha particles were scattered in this direction.

**Most** alpha particles were detected in this direction.

viewing lens

$\alpha$-particle source

A thin metal foil, e.g. gold, was bombarded with alpha particles in an evacuated chamber.

thin metal foil

**Observations**
- Most alpha particles travelled straight through with little or no deviation.
- About 1 particle in 8000 was deflected through a large angle.

**Conclusions**
- Most of matter is empty space, occupied by electrons of negligible mass.
- Most of the mass of an atom is contained in a very small positively charged space – the **nucleus**.

## SHOWING THE PATHS OF SCATTERED ALPHA PARTICLES

R

Alpha particles of the same energy approaching a nucleus

This particle **passes close to** the nucleus and is deflected through a large angle.

This particle **misses** the nucleus by a long way. The repulsive force is small so it deflects very little.

- Rutherford showed that the number of alpha particles deflected at different angles was as expected, assuming an inverse square law for the repulsion force between the nucleus and the alpha particles.

### FINDING THE RADIUS OF A NUCLEUS USING ALPHA PARTICLE SCATTERING

- The number of alpha particles scattered at each angle is compared with that expected, assuming an inverse square law.
- Alpha particles are reaching the nucleus when the actual and theoretical data do not agree.
- The **closest distance of approach** of the alpha particle to the nucleus gives an approximate value for the radius of the nucleus.
  (See 'Using electric fields', page 72.)

### MUST TAKE CARE

- Make sure that particles aimed nearer the nucleus
  - are **deflected most**
  - get **closer to the nucleus**.
- Make the tracks **symmetrical**. The incident and deflected tracks make the same angle with a radial line, e.g. R, drawn from the centre of the nucleus.

# NUCLEAR DENSITY AND STABILITY

## DENSITY OF NUCLEAR MATTER

- Data in this table show that $\dfrac{R}{A^{\frac{1}{3}}} \approx$ **constant**.

- For a wide range of nuclei:

  $R = r_0\, A^{\frac{1}{3}}$

  $r_0 \approx 1.20 \times 10^{-15}\,m$

  so $R = 1.20 \times 10^{-15}\, A^{\frac{1}{3}}\, m$

| Nucleus | $A$ | Measured radius/$10^{-15}$ m | $\dfrac{R}{A^{\frac{1}{3}}}/10^{-15}$ m |
|---|---|---|---|
| $^{115}_{49}$In | 115 | 5.80 | 1.19 |
| $^{122}_{51}$Sb | 122 | 5.97 | 1.21 |
| $^{197}_{79}$Au | 197 | 6.87 | 1.21 |
| $^{209}_{83}$Bi | 209 | 7.13 | 1.20 |

- Volume of a nucleus $= \dfrac{4}{3}\pi (r_0 A^{\frac{1}{3}})^3 = \dfrac{4}{3}\pi r_0^3 A$

- Mass of a nucleus $\approx A \times$ mass of a nucleon

  **Density of a nucleus** $\approx \dfrac{A \times \text{mass of a nucleon}}{\dfrac{4}{3}\pi r_0^3 A}$

  $\approx \dfrac{3 \times \textbf{mass of a nucleon}}{4\,\pi r_0^3}$

  > Proton mass $\approx$ neutron mass $\approx 1.7 \times 10^{-27}\,kg$
  >
  > Density of protons, neutrons and nuclei
  >
  > $\approx \dfrac{3 \times 1.7 \times 10^{-27}}{4 \times \pi \times (1.20 \times 10^{-15})^3}$
  >
  > $\approx 2.3 \times 10^{17}\,kg\,m^{-3}$

- The density does not depend on $A$, so is approximately constant.

## NUCLEAR STABILITY AND INSTABILITY

- The protons carry a positive charge and repel each other.

- The presence of the **strong force** between nucleons holds the nucleus together.

- The presence of neutrons separates the protons and reduces the repulsive effect of the protons.

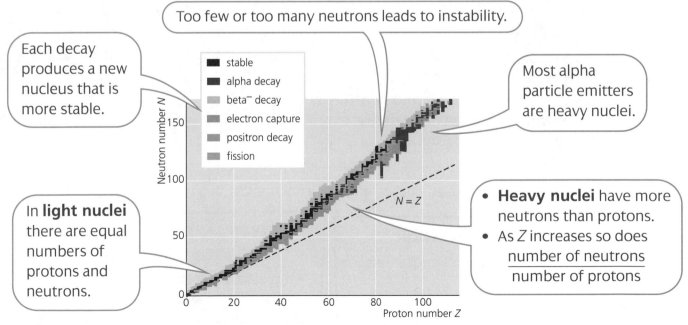

> Too few or too many neutrons leads to instability.

> Each decay produces a new nucleus that is more stable.

> Most alpha particle emitters are heavy nuclei.

> In **light nuclei** there are equal numbers of protons and neutrons.

legend: stable, alpha decay, beta⁻ decay, electron capture, positron decay, fission

$N = Z$

Neutron number $N$ — Proton number $Z$

> - **Heavy nuclei** have more neutrons than protons.
> - As $Z$ increases so does $\dfrac{\text{number of neutrons}}{\text{number of protons}}$

- An unstable nucleus can become more stable by:
  - emission of alpha, beta$^+$, beta$^-$ or gamma radiation
  - electron capture.

- Before a nucleus reaches a stable structure many transformations may take place.

- There are four main **radioactive decay series** (or **decay chains**):
  Thorium series ($^{232}_{90}$Th), Neptunium series ($^{237}_{93}$Np), Uranium series ($^{238}_{92}$U), Actinium series ($^{235}_{92}$U).

- The nuclide in brackets is the first in the chain.

# RADIOACTIVE DECAY

- **Radioactive decay** is:
  - spontaneous – the nucleus decays without the need for any external stimulus
  - random – there is no way of knowing when a particular nucleus will decay.

- The nucleus that decays is called the **parent nucleus**.

- The nucleus that is formed as a result of the decay is called the **daughter nucleus**.

## ALPHA (α) DECAY

- An α **particle** is a helium-4 ($^4_2$He) nucleus.

- When a nucleus emits an alpha particle:
  - the $Z$ number decreases by 2
  - the $A$ number decreases by 4.

- An example of α-decay is
$$^{220}_{86}\text{Rn} \Rightarrow {}^{216}_{84}\text{Po} + {}^4_2\alpha + 1.0 \times 10^{-12} \text{ J}$$

## MUST REMEMBER

In all **nuclear equations**:

- The sum of the $Z$ numbers on each side of the equation must be equal (to **conserve charge**).

- The sum of the $A$ numbers on each side must be equal (to conserve **baryon number**).

- This is the sum of the kinetic energies of the daughter nucleus and the alpha particle.
- The daughter nucleus recoils in the opposite direction to the alpha particle to conserve momentum.

## BETA-MINUS (β⁻) DECAY

- A β⁻ particle is a fast-moving **electron.**

- When a nucleus emits a beta⁻ particle:
  - the $Z$ number increases by 1
  - the $A$ number is unchanged.

- An example of β⁻-decay is
$$^{40}_{19}\text{K} \Rightarrow {}^{40}_{20}\text{Ca} + {}^{\ 0}_{-1}\beta^- + {}^0_0\overline{\nu} + 2.2 \times 10^{-13} \text{ J}$$

## BETA-PLUS (β⁺) DECAY

- A β⁺ particle is a **positron.**

- When a nucleus emits a beta⁺ particle:
  - the $Z$ number decreases by 1
  - the $A$ number is unchanged.

- An example of β⁺-decay is
$$^{25}_{13}\text{Al} \Rightarrow {}^{25}_{12}\text{Mg} + {}^0_1\beta^+ + {}^0_0\nu + 5.1 \times 10^{-13} \text{ J}$$

- To conserve **lepton number** and **momentum**:
  - an **anti-neutrino** is emitted in β⁻ **decay**
  - a **neutrino** is emitted in β⁺ **decay.**

- The energy released in the decay is shared between the three particles.

- Beta particles emitted by a given nuclide may have energies from 0 to the maximum possible.

## EMISSION OF GAMMA (γ) RADIATION

- Following α or β decay, a nucleus may end up in an excited state.

- **Nuclear excited states** are well-defined states like those of the atomic electrons.

- The nucleus becomes more stable by emitting a **gamma ray photon**.

- The energy levels are further apart in nuclei so the photon energies are much larger.

- When a nucleus emits a gamma ray photon the $Z$ and $A$ numbers are both unchanged.

## ELECTRON CAPTURE

- This is another way of achieving stability.

- An electron in an inner shell of the atom is captured by the nucleus.

- When electron capture takes place:
  - the $Z$ number decreases by 1
  - the $A$ number is unchanged.

- An example of electron capture is:
$$^{64}_{29}\text{Cu} + {}^{\ 0}_{-1}\text{e} \Rightarrow {}^{64}_{28}\text{Ni} + {}^0_0\nu + 2.7 \times 10^{-13} \text{ J}$$

- The emitted neutrino conserves lepton number and momentum and takes some of the energy released.

# PROPERTIES OF α, β AND γ RADIATION

## IONISING ABILITY

- α, β and γ radiation can ionise atoms and molecules.
- The radiation **loses energy** by ionisation as it passes through matter.

- **Ionisation** is the process of removing one or more electrons from an atom or molecule.
- The atom or molecule becomes a **positively charged ion**.
- To ionise air (nitrogen or oxygen molecules) requires about 28 eV of energy ($4.5 \times 10^{-18}$ J).

- A typical α or β particle has an energy between 1 MeV and 5 MeV.
- The energy of γ ray photons range from about 10 keV to about 1 MeV. ($1\,eV = 1.6 \times 10^{-19}$ J)
- Each particle can ionise many atoms or molecules.

## ABSORPTION PROPERTIES

**Alpha particles:**
- are large so collide frequently with atoms
- lose all their energy over a short distance
- travel 1 to 5 cm in air
- are absorbed by a thin sheet of paper or card.

- A source **outside** a body damages only the skin, as α particles cannot penetrate further.
- A source **inside** the body can do much damage to the soft tissue.

**Beta particles:**
- produce ions less easily
- travel up to 100 cm before losing all their energy
- can be stopped by a sheet of aluminium 1–2 mm thick.

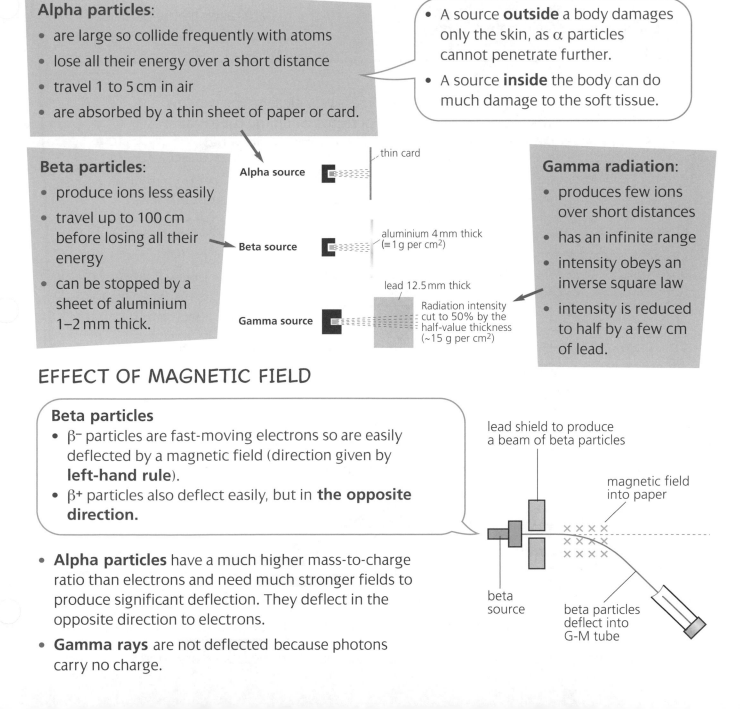

thin card

Alpha source

aluminium 4 mm thick ($\equiv$ 1 g per cm²)

Beta source

lead 12.5 mm thick

Gamma source

Radiation intensity cut to 50% by the half-value thickness (~15 g per cm²)

**Gamma radiation:**
- produces few ions over short distances
- has an infinite range
- intensity obeys an inverse square law
- intensity is reduced to half by a few cm of lead.

## EFFECT OF MAGNETIC FIELD

**Beta particles**
- β⁻ particles are fast-moving electrons so are easily deflected by a magnetic field (direction given by **left-hand rule**).
- β⁺ particles also deflect easily, but in **the opposite direction.**

lead shield to produce a beam of beta particles

magnetic field into paper

beta source

beta particles deflect into G-M tube

- **Alpha particles** have a much higher mass-to-charge ratio than electrons and need much stronger fields to produce significant deflection. They deflect in the opposite direction to electrons.
- **Gamma rays** are not deflected because photons carry no charge.

# DETECTION OF RADIATION

- **Radiation** is detected by the ionisation that it produces.

## PHOTOGRAPHIC PLATE

- All types of radiation can be detected using a **photographic plate**.
- The degree of blackening is a measure of the **exposure** to radiation.
- Workers wear badges fitted with film to enable measurement of their exposure to radiation.
- The badges have absorbers of different materials and thickness in front of the film so that different parts of the film detect different types of radiation.

- An **alpha particle track** is straight with a high density of liquid droplets.

- A **beta particle track** is a straight track when it is emitted, but the track becomes more straggled as it loses energy and slows down. The track is not as dense as an alpha particle track as fewer ions are produced.

- A **gamma ray track** is very faint as it produces relatively few ions in a short distance.

## CLOUD CHAMBER

- Radiation produces ions as it passes through a super-cooled vapour.
- Droplets of liquid form on the ions produced.
- The droplets of vapour show the path taken by the particle.

## GEIGER-COUNTER

- Some alpha particles are detected if the source is very close to the window.
- All beta particles that enter the tube will be detected when the tube is 'active'.
- Less than 1% of gamma ray photons that enter the tube are detected.

- Ions are produced when the particle or gamma ray photon enters the tube.
- A high electric field produces many more ions through an **avalanche effect** and the tube quickly becomes conducting.

Thin window so that particles can enter the tube.

Each detected particle produces a pulse of current that is recorded by an electronic counter.

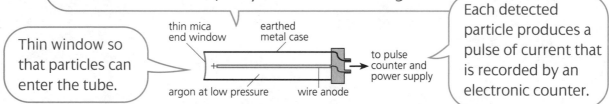

thin mica end window · earthed metal case · to pulse counter and power supply · argon at low pressure · wire anode

- After a particle is detected, it takes time for the tube to be ready to detect another.
- The recovery time is called the **dead time**. Corrections can be applied to find the true **count rate**.

# BACKGROUND RADIATION

- Background radiation is the radiation that occurs naturally.

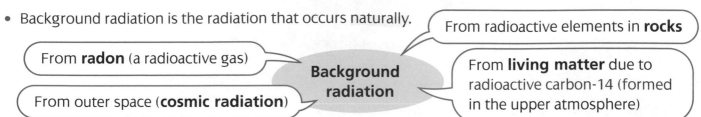

From radioactive elements in **rocks**

From **radon** (a radioactive gas)

**Background radiation**

From **living matter** due to radioactive carbon-14 (formed in the upper atmosphere)

From outer space (**cosmic radiation**)

- To find the count rate due to a source in an experiment, a correction has to be applied.

**Corrected count rate = measured count rate − background count rate**

- The **background count rate** is the count rate measured when the laboratory source is not present.

# EXPERIMENTS WITH RADIOACTIVITY

## ABSORPTION EXPERIMENTS

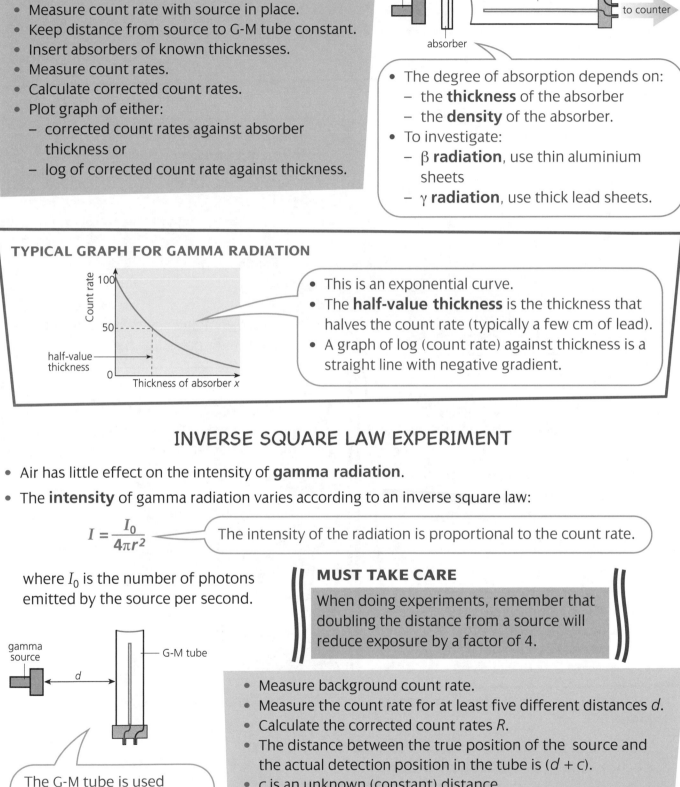

- Measure background count rate.
- Measure count rate with source in place.
- Keep distance from source to G-M tube constant.
- Insert absorbers of known thicknesses.
- Measure count rates.
- Calculate corrected count rates.
- Plot graph of either:
  - corrected count rates against absorber thickness or
  - log of corrected count rate against thickness.

- The degree of absorption depends on:
  - the **thickness** of the absorber
  - the **density** of the absorber.
- To investigate:
  - β **radiation**, use thin aluminium sheets
  - γ **radiation**, use thick lead sheets.

### TYPICAL GRAPH FOR GAMMA RADIATION

- This is an exponential curve.
- The **half-value thickness** is the thickness that halves the count rate (typically a few cm of lead).
- A graph of log (count rate) against thickness is a straight line with negative gradient.

## INVERSE SQUARE LAW EXPERIMENT

- Air has little effect on the intensity of **gamma radiation**.

- The **intensity** of gamma radiation varies according to an inverse square law:

$$I = \frac{I_0}{4\pi r^2}$$

The intensity of the radiation is proportional to the count rate.

where $I_0$ is the number of photons emitted by the source per second.

**MUST TAKE CARE**

When doing experiments, remember that doubling the distance from a source will reduce exposure by a factor of 4.

The G-M tube is used sideways so that **only gamma radiation** is detected. (Any alpha and beta radiation is absorbed by the outer metal electrode and case).

- Measure background count rate.
- Measure the count rate for at least five different distances $d$.
- Calculate the corrected count rates $R$.
- The distance between the true position of the source and the actual detection position in the tube is $(d + c)$.
- $c$ is an unknown (constant) distance.
- For an inverse square law:

$$\text{corrected count rate } R \propto \frac{1}{(d + c)^2}$$

$$d + c \propto \frac{1}{\sqrt{R}} \Rightarrow d = \frac{k}{\sqrt{R}} - c$$

- The inverse square law is shown by a straight line graph for $d$ against $\frac{1}{\sqrt{R}}$

# DECAY CONSTANT AND HALF-LIFE

- Radioactive decay is a random process.
- Every atom of a given radioactive nuclide has the same probability of decay in each one second time interval.
- It is not possible to tell when any given atom will disintegrate.
- Provided that there is a large number of radioactive atoms, the proportion of atoms that decay in each second is a constant.

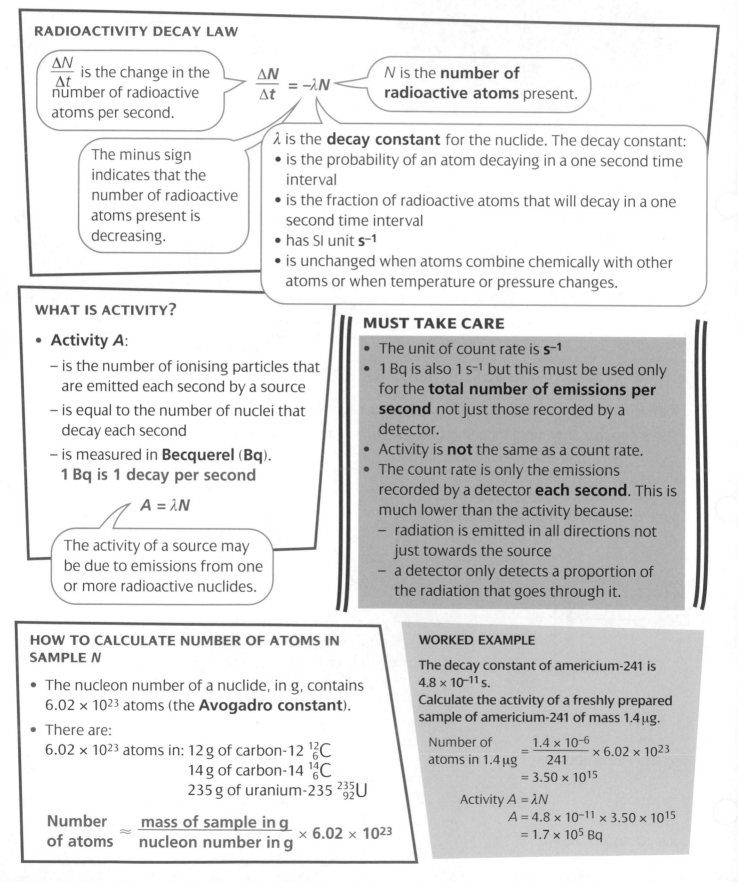

## RADIOACTIVITY DECAY LAW

$\dfrac{\Delta N}{\Delta t}$ is the change in the number of radioactive atoms per second.

$$\frac{\Delta N}{\Delta t} = -\lambda N$$

$N$ is the **number of radioactive atoms** present.

The minus sign indicates that the number of radioactive atoms present is decreasing.

$\lambda$ is the **decay constant** for the nuclide. The decay constant:
- is the probability of an atom decaying in a one second time interval
- is the fraction of radioactive atoms that will decay in a one second time interval
- has SI unit **s⁻¹**
- is unchanged when atoms combine chemically with other atoms or when temperature or pressure changes.

## WHAT IS ACTIVITY?

- **Activity $A$:**
  - is the number of ionising particles that are emitted each second by a source
  - is equal to the number of nuclei that decay each second
  - is measured in **Becquerel (Bq)**.
    **1 Bq is 1 decay per second**

  $$A = \lambda N$$

The activity of a source may be due to emissions from one or more radioactive nuclides.

## MUST TAKE CARE

- The unit of count rate is **s⁻¹**
- 1 Bq is also 1 s⁻¹ but this must be used only for the **total number of emissions per second** not just those recorded by a detector.
- Activity is **not** the same as a count rate.
- The count rate is only the emissions recorded by a detector **each second**. This is much lower than the activity because:
  - radiation is emitted in all directions not just towards the source
  - a detector only detects a proportion of the radiation that goes through it.

## HOW TO CALCULATE NUMBER OF ATOMS IN SAMPLE $N$

- The nucleon number of a nuclide, in g, contains $6.02 \times 10^{23}$ atoms (the **Avogadro constant**).
- There are:
  $6.02 \times 10^{23}$ atoms in: 12 g of carbon-12 $^{12}_{6}C$
  14 g of carbon-14 $^{14}_{6}C$
  235 g of uranium-235 $^{235}_{92}U$

$$\text{Number of atoms} \approx \frac{\text{mass of sample in g}}{\text{nucleon number in g}} \times 6.02 \times 10^{23}$$

### WORKED EXAMPLE

The decay constant of americium-241 is $4.8 \times 10^{-11}$ s.
Calculate the activity of a freshly prepared sample of americium-241 of mass 1.4 μg.

$$\text{Number of atoms in 1.4 μg} = \frac{1.4 \times 10^{-6}}{241} \times 6.02 \times 10^{23}$$
$$= 3.50 \times 10^{15}$$

Activity $A = \lambda N$
$$A = 4.8 \times 10^{-11} \times 3.50 \times 10^{15}$$
$$= 1.7 \times 10^{5} \text{ Bq}$$

# HALF-LIFE

- Because the rate of decay is proportional to the number of radioactive nuclei, a radioactive nuclide decays exponentially.

**Half-life $T_{\frac{1}{2}}$ of a radioactive nuclide is the mean time taken for the number of radioactive nuclei of the nuclide to halve.**

- This graph is for a sample of radon gas.
- The count rate takes 55 s to fall from 400 to 200 and from 200 to 100.
- The half-life of radon is 55 s.

- Random decay leads to some count rates being higher than expected and some lower.
- The randomness is most noticeable at low count rates when there are fewer atoms decaying.

**MUST TAKE CARE**

When using a graph to find half-life, take the measurement for **at least two** different starting values of count rate.

**MUST REMEMBER**

**Count rate ∝ activity $A$ ∝ number of radioactive nuclei $N$**

So a graph of $A$ against time and a graph of $N$ against time have the same shape as a graph of count rate against time.

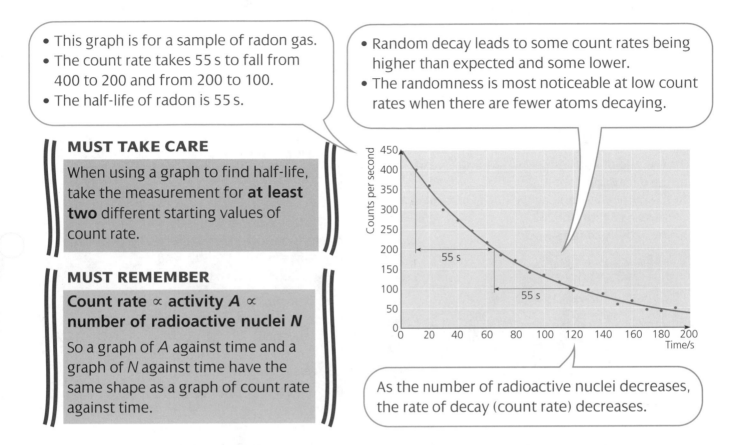

As the number of radioactive nuclei decreases, the rate of decay (count rate) decreases.

## RADIOACTIVE DECAY EQUATIONS

- These show how activity and number of radioactive (R/A) nuclei present vary with time.

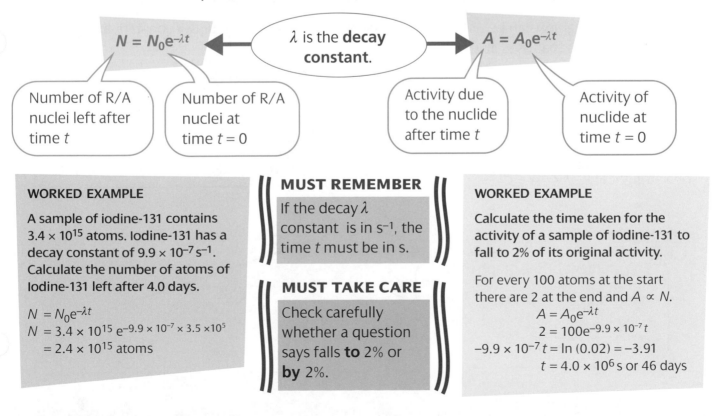

$$N = N_0 e^{-\lambda t}$$

$\lambda$ is the **decay constant.**

$$A = A_0 e^{-\lambda t}$$

Number of R/A nuclei left after time $t$

Number of R/A nuclei at time $t = 0$

Activity due to the nuclide after time $t$

Activity of nuclide at time $t = 0$

**WORKED EXAMPLE**

A sample of iodine-131 contains $3.4 \times 10^{15}$ atoms. Iodine-131 has a decay constant of $9.9 \times 10^{-7}\,s^{-1}$. Calculate the number of atoms of Iodine-131 left after 4.0 days.

$N = N_0 e^{-\lambda t}$
$N = 3.4 \times 10^{15}\, e^{-9.9 \times 10^{-7} \times 3.5 \times 10^5}$
$\quad = 2.4 \times 10^{15}$ atoms

**MUST REMEMBER**

If the decay $\lambda$ constant is in $s^{-1}$, the time $t$ must be in s.

**MUST TAKE CARE**

Check carefully whether a question says falls **to** 2% or **by** 2%.

**WORKED EXAMPLE**

Calculate the time taken for the activity of a sample of iodine-131 to fall to 2% of its original activity.

For every 100 atoms at the start there are 2 at the end and $A \propto N$.
$$A = A_0 e^{-\lambda t}$$
$$2 = 100 e^{-9.9 \times 10^{-7} t}$$
$$-9.9 \times 10^{-7} t = \ln(0.02) = -3.91$$
$$t = 4.0 \times 10^6\,s \text{ or } 46 \text{ days}$$

# RELATING DECAY CONSTANT TO HALF-LIFE

$$N = N_0 e^{-\lambda t}$$

- In one half-life $T_{\frac{1}{2}}$ the number of radioactive nuclei changes from $N_0$ to $\frac{N_0}{2}$

$$\frac{N_0}{2} = N_0 e^{-\lambda t}$$

$$\ln(\frac{1}{2}) = -0.693 = -\lambda T_{\frac{1}{2}}$$

$$T_{\frac{1}{2}} = \frac{0.693}{\lambda}$$

> It is more usual to quote the **half-life** of a radioactive nuclide rather than decay constant.

> A **long** half-life means a **small** decay constant and vice versa.

**WORKED EXAMPLES**

1 The half-life of thorium-232 is $1.41 \times 10^{10}$ years. Calculate:
  (a) the decay constant in $s^{-1}$
  (b) the number of radioactive nuclei present in a 35 mg sample of thorium-232
  (c) the activity of the sample.

(a) 1 year $= 3.15 \times 10^7$ s

$$\lambda = \frac{0.693}{1.41 \times 10^{10} \times 3.15 \times 10^7} = 1.56 \times 10^{-18}\,s^{-1}$$

(b) Number of atoms in sample $= \dfrac{35 \times 10^{-3}}{232} \times 6.02 \times 10^{23}$

$$= 9.08 \times 10^{19}$$

(c) Activity $= \lambda N = 1.56 \times 10^{-18} \times 9.08 \times 10^{19}$

$$= 142\,Bq$$

2 How many alpha and beta$^-$ particles are emitted when one thorium-232 ($^{232}_{90}$Th) nucleus decays to lead-212 ($^{212}_{82}$Pb)?

Nucleon number $N$ falls by 4 for each $\alpha$ emission and does not change for $\beta^-$ emission.
Change in $N = 232 - 212 = 20$
Number of alpha particles emitted $= 5$

Proton number $Z$ decreases by 2 for each $\alpha$ emission and increases by 1 for each $\beta^-$ emission.
Decrease in $Z$ due to alpha emission $= 10$
Overall decrease in $Z = 90 - 82 = 8$
Number of beta$^-$ particles emitted $= 2$

# LOGARITHMIC GRAPH OF RADIOACTIVE DECAY

- For exponential decay, a graph of ln $N$ against time is a straight line with negative gradient.
- Graphs of ln $A$ or ln (count rate) against time are also straight-line graphs.
- The straight-line graph can be used to determine a mean value for the decay constant and hence the half-life that uses all the data.
- The log graph is preferred to the curve because it is easier to draw the best-fit line.
- Radioactive decay equation is: $N = N_0 e^{-\lambda t}$
  Taking ln of both sides gives: $\ln N = \ln N_0 - \lambda t$

$$\ln N = -\lambda t + \ln N_0$$

  Compare this with the equation of a straight line: $y = mx + c$

> This intercept is ln $N_0$
> ln $N_0 \approx 9.2$
> Original number of atoms present $N_0 \approx 9900$

> The gradient of the graph $m$ is $-\lambda$
> $$-\lambda \approx \frac{6.6 - 9.2}{200} = -0.013$$
> $$T_{\frac{1}{2}} = \frac{0.693}{\lambda} \approx \frac{0.69}{0.013} = 53\,s$$

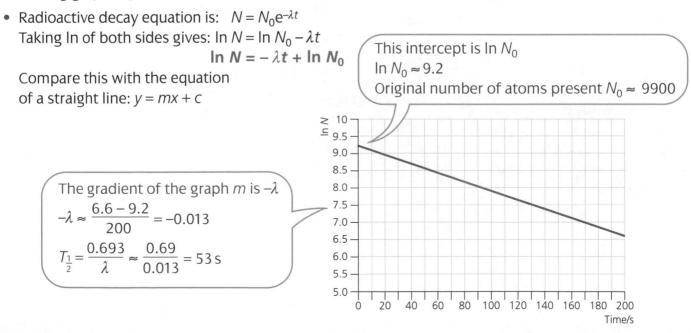

# MASS–ENERGY EQUIVALENCE

The mass of a system of particles is increased when energy is supplied to it.

Einstein's mass–energy relationship states
$$E = mc^2$$
or
$$\Delta E = c^2 \Delta m$$
where $c$ is the speed of e-m radiation.

When energy is lost by a system of particles, the mass of the system decreases.

## MASS DEFECT

- The mass of a nucleus consisting of $Z$ protons and $(A - Z)$ neutrons has a lower mass than the sum of the masses of protons and neutrons in it.

  Mass of nucleus $< Zm_p + (A - Z)m_n$
  where $m_p$ = proton mass
  $m_n$ = neutron mass

- The difference between the total mass of the individual particles and the mass of the nucleus is the **mass defect**.

## BINDING ENERGY

- To split up a nucleus into its individual protons and neutrons, energy would have to be supplied to it. This energy is called the **binding energy**.

- The same amount of energy would appear in another form when the nucleus is made.

- The total binding energy of a nucleus is equal to the energy equivalence of the mass defect.

---

- Binding energies are usually given as the **binding energy per nucleon** (BEPN) in MeV (1 MeV = $1.602 \times 10^{-13}$ J).
- The total binding energy of a nucleus ($BE_{total}$) is given by:

  $$BE_{total} = \text{Nucleon number } A \times BEPN$$

---

**WORKED EXAMPLE**

Calculate the mass defect (in kg), the binding energy (in J), and the binding energy per nucleon (in J per nucleon and in MeV per nucleon) for a nucleus of helium-4 ($^4_2He$) and a nucleus of uranium-235 ($^{235}_{92}U$)

Mass of proton $m_p = 1.6740 \times 10^{-27}$ kg
Mass of neutron $m_n = 1.6754 \times 10^{-27}$ kg
Mass of helium-4 nucleus = $6.6483 \times 10^{-27}$ kg
Mass of uranium-235 nucleus = $390.4080 \times 10^{-27}$ kg
Speed of electromagnetic radiation $c = 2.998 \times 10^8$ m s$^{-1}$

**Helium**

$$
\begin{aligned}
\text{Mass of protons} &= 2 \times 1.6740 \times 10^{-27} \\
&= 3.3480 \times 10^{-27} \text{ kg} \\
\text{Mass of neutrons} &= (4 - 2) \times 1.6754 \times 10^{-27} \\
&= 3.3508 \times 10^{-27} \text{ kg} \\
\text{Mass defect} &= (\text{mass of 2 protons + mass of 2 neutrons}) \\
&\quad - (\text{mass of helium nucleus}) \\
&= 6.6988 \times 10^{-27} - 6.6483 \times 10^{-27} \\
&= 0.0505 \times 10^{-27} \text{ kg}
\end{aligned}
$$

$$
\begin{aligned}
\text{Total binding energy} &= mc^2 \\
&= 0.0505 \times 10^{-27} \times (2.998 \times 10^8)^2 \\
&= 4.5389 \times 10^{-12} \text{ J}
\end{aligned}
$$

$$
\begin{aligned}
\text{Binding energy per nucleon} &= \frac{4.5389 \times 10^{-12}}{4} \\
&= 1.1347 \times 10^{-12} \text{ J}
\end{aligned}
$$

$$
\begin{aligned}
\text{Binding energy per nucleon} &= \frac{1.1347 \times 10^{-12}}{1.602 \times 10^{-13}} \\
&= 7.08 \text{ MeV}
\end{aligned}
$$

**Uranium**

$$
\begin{aligned}
\text{Mass of protons} &= 92 \times 1.6740 \times 10^{-27} \\
&= 154.008 \times 10^{-27} \text{ kg} \\
\text{Mass of neutrons} &= (235 - 92) \times 1.6754 \times 10^{-27} \\
&= 239.582 \times 10^{-27} \text{ kg} \\
\text{Mass defect} &= 393.590 \times 10^{-27} - 390.408 \times 10^{-27} \\
&= 3.182 \times 10^{-27} \text{ kg}
\end{aligned}
$$

$$
\begin{aligned}
\text{Total binding energy} &= mc^2 \\
&= 3.182 \times 10^{-27} \times (2.998 \times 10^8)^2 \\
&= 2.860 \times 10^{-10} \text{ J}
\end{aligned}
$$

$$
\begin{aligned}
\text{Binding energy per nucleon} &= \frac{2.760 \times 10^{-10}}{235} \\
&= 1.217 \times 10^{-12} \text{ J}
\end{aligned}
$$

$$
\begin{aligned}
\text{Binding energy per nucleon} &= \frac{1.217 \times 10^{-12}}{1.602 \times 10^{-13}} \\
&= 7.60 \text{ MeV}
\end{aligned}
$$

# BINDING ENERGY CURVE

- This graph gives the binding energies **per nucleon**.
- Remember that this is the energy **input** needed to remove one nucleon from the nucleus.

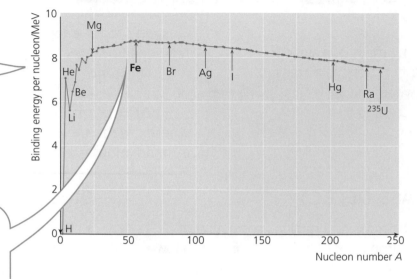

More energy is needed to remove a nucleon from iron than any other nucleus. It is the most stable nucleus.

- The curve can be used to show that energy is released when:
  - two light nucleons combine to form a heavier nucleus (see 'Nuclear fusion', page 110)
  - a heavy nucleus splits into lighter nuclei (see 'Nuclear fission', page 109).

---

## CALCULATING ENERGY RELEASED IN ALPHA DECAY

- A nucleus can only decay spontaneously if:
  - its mass is greater than the mass of an alpha particle and the remaining nucleus (the daughter nucleus)
  - there is an increase in the total binding energy as a result of the decay
- The energy released appears as kinetic energy of the daughter and the alpha particle.

---

**WORKED EXAMPLE**

Calculate the total kinetic energy of the decay products when uranium-238 (BEPN = 7.57 MeV) decays to thorium-234 (BEPN = 7.60 MeV) by emitting an alpha particle (BEPN = 7.08 MeV).

Total binding energy of uranium-238 $= 238 \times 7.57$ MeV
$= 1801.7$ MeV

Total binding energy of thorium-234 $= 234 \times 7.60$ MeV
$= 1778.4$ MeV

Total binding energy of alpha particle $= 4 \times 7.08$ MeV
$= 28.3$ MeV

Total binding energy before emission $= 1801.7$ MeV
Total binding energy after emission $= 1778.4 + 28.3$
$= 1806.7$ MeV

Increase in binding energy = K.E. of alpha particle and recoil nucleus
$= 5.0$ MeV

**MUST REMEMBER**

Energy is released when there is a higher value for binding energy.

To conserve momentum, the velocity of the recoil nucleus is **much lower** than that of the alpha particle (about 1/59). This results in the alpha particle taking most (about 98%) of the kinetic energy.

# NUCLEAR FISSION AND FUSION

## NUCLEAR FISSION

- Some nuclei can be made to split into two smaller nuclei following the addition of a neutron to the nucleus. This is **neutron-induced fission**.

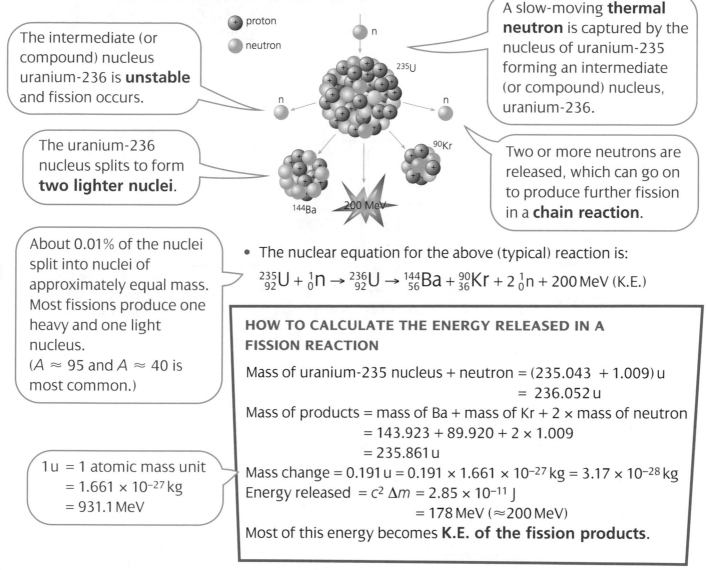

The intermediate (or compound) nucleus uranium-236 is **unstable** and fission occurs.

A slow-moving **thermal neutron** is captured by the nucleus of uranium-235 forming an intermediate (or compound) nucleus, uranium-236.

The uranium-236 nucleus splits to form **two lighter nuclei**.

Two or more neutrons are released, which can go on to produce further fission in a **chain reaction**.

About 0.01% of the nuclei split into nuclei of approximately equal mass. Most fissions produce one heavy and one light nucleus.
($A \approx 95$ and $A \approx 40$ is most common.)

- The nuclear equation for the above (typical) reaction is:

$$^{235}_{92}U + {}^{1}_{0}n \rightarrow {}^{236}_{92}U \rightarrow {}^{144}_{56}Ba + {}^{90}_{36}Kr + 2\,{}^{1}_{0}n + 200\,MeV\ (K.E.)$$

**HOW TO CALCULATE THE ENERGY RELEASED IN A FISSION REACTION**

Mass of uranium-235 nucleus + neutron = $(235.043 + 1.009)\,u$
$= 236.052\,u$

Mass of products = mass of Ba + mass of Kr + 2 × mass of neutron
$= 143.923 + 89.920 + 2 \times 1.009$
$= 235.861\,u$

Mass change = $0.191\,u = 0.191 \times 1.661 \times 10^{-27}\,kg = 3.17 \times 10^{-28}\,kg$
Energy released = $c^2\,\Delta m = 2.85 \times 10^{-11}\,J$
$= 178\,MeV\ (\approx 200\,MeV)$

Most of this energy becomes **K.E. of the fission products**.

$1\,u = 1$ atomic mass unit
$= 1.661 \times 10^{-27}\,kg$
$= 931.1\,MeV$

## CRITICAL MASS

- The **critical mass** is the mass of a lump of fissile material in which a chain reaction can be sustained without the need to add neutrons from an outside source.

- The critical mass depends on the shape of the lump of fissile material.

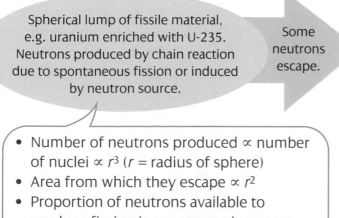

Spherical lump of fissile material, e.g. uranium enriched with U-235. Neutrons produced by chain reaction due to spontaneous fission or induced by neutron source.

Some neutrons escape.

- Number of neutrons produced $\propto$ number of nuclei $\propto r^3$ ($r$ = radius of sphere)
- Area from which they escape $\propto r^2$
- Proportion of neutrons available to produce fission increases as $r$ increases.

- **Mass of sphere < critical mass**
  More neutrons escape per second than are produced $\Rightarrow$ reaction not self-sustaining

- **Mass of sphere = critical mass**
  Neutrons produced each second = number escaping $\Rightarrow$ reaction just self-sustaining

- **Mass of sphere > critical mass**
  More neutrons produced each second than number escaping $\Rightarrow$ number of fission reactions increases with time $\Rightarrow$ material becomes hotter $\Rightarrow$ material ultimately melts

# FISSION REACTOR

- The diagram shows a pressurised water reactor (PWR).

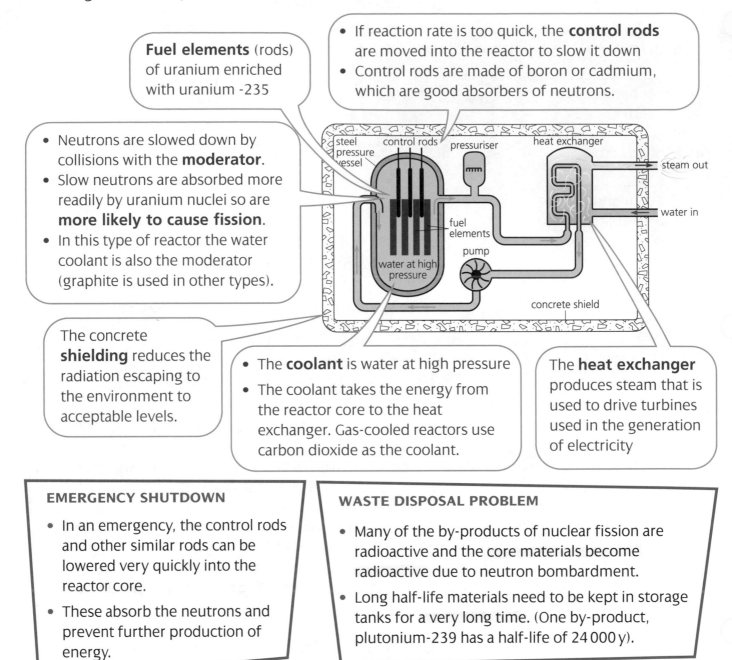

**Fuel elements** (rods) of uranium enriched with uranium -235

- If reaction rate is too quick, the **control rods** are moved into the reactor to slow it down
- Control rods are made of boron or cadmium, which are good absorbers of neutrons.

- Neutrons are slowed down by collisions with the **moderator**.
- Slow neutrons are absorbed more readily by uranium nuclei so are **more likely to cause fission**.
- In this type of reactor the water coolant is also the moderator (graphite is used in other types).

The concrete **shielding** reduces the radiation escaping to the environment to acceptable levels.

- The **coolant** is water at high pressure
- The coolant takes the energy from the reactor core to the heat exchanger. Gas-cooled reactors use carbon dioxide as the coolant.

The **heat exchanger** produces steam that is used to drive turbines used in the generation of electricity

Diagram labels: steel pressure vessel, control rods, pressuriser, heat exchanger, steam out, water in, fuel elements, pump, water at high pressure, concrete shield

## EMERGENCY SHUTDOWN

- In an emergency, the control rods and other similar rods can be lowered very quickly into the reactor core.
- These absorb the neutrons and prevent further production of energy.

## WASTE DISPOSAL PROBLEM

- Many of the by-products of nuclear fission are radioactive and the core materials become radioactive due to neutron bombardment.
- Long half-life materials need to be kept in storage tanks for a very long time. (One by-product, plutonium-239 has a half-life of 24 000 y).

# NUCLEAR FUSION

- When two light nuclei fuse together, the binding energy increases and energy is released.
- Nuclear fusion produces energy in stars (see Stellar evolution, pages 126–128)

**WORKED EXAMPLE**

Calculate the energy released when two deuterium ($^2_1H$) nuclei fuse together to form helium-4.
Binding energy of deuterium = 1.1 MeV per nucleon
Binding energy of helium-4 = 7.1 MeV per nucleon

Binding energy of each deuteron = 2 × 1.1 = 2.2 MeV
Initial total binding energy = 2 × 2.2 = 4.4 MeV
Binding energy of helium = 4 × 7.1 = 28.4 MeV
Increase in binding energy = 28.4 − 4.4 = 24.0 MeV
The increase in binding energy = the energy released

If helium could be formed from four hydrogen nuclei, 4 × 7.1 = 28.4 MeV would be available for each helium nucleus formed.

Although the energy released per fusion is about **ten times lower** than the energy released per fission, the raw materials (hydrogen and deuterium) are more plentiful than uranium.

# FUSION (JET) REACTOR

- The **J**oint **E**uropean **T**orus reactor shown is an experimental 'tokomak' fusion reactor.

The temperature of the plasma is **raised** by:

- inducing currents in the plasma using alternating currents in these coils
- using high powered lasers.

At high temperatures, electrons are separated from the nuclei and a **plasma** is formed. Fusion of the nuclei in the plasma releases energy.

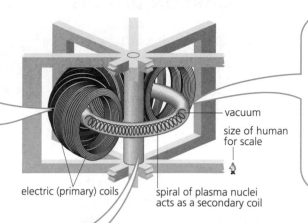

vacuum

size of human for scale

electric (primary) coils

spiral of plasma nuclei acts as a secondary coil

- Because of the high temperature, the JET reactor contains the plasma using a magnetic field.
- The particles move in spiral paths in a ring called the **torus**.

- A major problem in the design of reactors is how to contain the plasma for a sufficiently long time.
- In stars, the plasma is contained using gravitational forces.
- Inertial confinement is being tried in which intense laser beams are used to keep the plasma in place.

## DEUTERIUM-TRITIUM REACTION – THE LIKELY REACTION

- The positively-charged nuclei repel so that as they approach, their K.E. decreases and P.E. increases.
- To fuse, a temperature of at least $1 \times 10^8$ K is needed.

then

- neutron
- proton
- deuterium
- tritium
- helium

Two deuterons (deuterium nuclei) in the plasma combine to form **a tritium nucleus and a proton**.

A tritium nucleus and a deuteron combine to form **a helium nucleus and a neutron**.

- Nuclear equations for deuterium-tritium reactions:

$$^2_1H + {}^2_1H \rightarrow {}^3_1H + {}^1_1H + \text{energy}$$

$$^3_1H + {}^2_1H \rightarrow {}^4_2He + {}^1_0n + \text{energy}$$

- The net effect is:
  - the formation of a helium nucleus from two deuterons
  - and the splitting of a deuteron into a proton and a neutron.
- 21.6 MeV of energy is released from the fusion of each pair of deuterons.

### HOW COULD ENERGY BE EXTRACTED FROM A FUSION REACTOR?

#### Possible technique

- Reactor is surrounded with a **lithium 'blanket'**.
- Lithium melts due to energy absorbed from neutrons produced in the reactor.
- Molten lithium passes to heat exchanger, which produces steam.
- Steam is used to drive turbines and produce electricity.
- Tritium, produced in reaction between neutrons and lithium, is extracted for use in the reactor.

# USING RADIOACTIVITY

## CARBON DATING

Neutrons are produced by cosmic rays in the upper atmosphere. → Neutrons collide with nitrogen nuclei producing radioactive carbon-14

$$^{14}_{7}N + ^{1}_{0}n \rightarrow ^{14}_{6}C + ^{1}_{1}p$$

→ Atmosphere contains mostly carbon-12 but has traces of radioactive carbon-14.

The radiocarbon present now decays with a half-life of 5700 y. This can be used to determine the approximate date when the plant or animal died (or the age of objects made from it). ← When the plant or animal dies, it stops taking in atmospheric carbon. ← The isotopes of carbon behave the same chemically, and are used by plants and animals to grow.

The activity $A$ per kg of carbon taken from **old artefacts** made from dead matter, e.g. trees, is measured.

- To find the age, use:
  $$A = A_0 e^{-\lambda t}$$
- Accuracy is limited by the assumption that the proportion of carbon in the atmosphere is unchanged over time.

The activity per kg of carbon taken from **living matter** is measured to give $A_0$

## MEDICAL USES OF RADIOACTIVITY

### USE AS TRACERS

- Radioactive isotopes are used by the body in the same way as stable isotopes.
  - Chemicals can be 'labelled' with a small known proportion of radioactive nuclei.
  - The uptake of the element by any part of the body can be monitored by a detector outside the body, e.g. iodine can be labelled with iodine-132 to investigate the functioning of the thyroid.

- Suitable isotopes:
  - have a short half-life
  - emit radiation that can be monitored by detectors outside the body
  - produce as little ionisation in the body as possible (gamma is best but beta is also used).

- **Technetium-99M** is commonly used – a **gamma emitter** with a **half-life** of 6 h and decays to ordinary technetium which emits very little radiation.

### USE AS TREATMENT (THERAPY)

- Ionising radiation is used to treat cancer because it can kill malignant cells.
- Care is needed in treatment because healthy tissue can be damaged too.

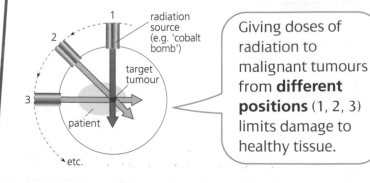

Giving doses of radiation to malignant tumours from **different positions** (1, 2, 3) limits damage to healthy tissue.

- Needles containing an alpha source can be inserted in the body to treat tumours near the outside of the body.
- Radioactive nuclides can be injected or given in a drink, e.g. gold-198, to treat malignant tissue in the abdomen.

# PARTICLE PHYSICS

## CLASSIFYING PARTICLES

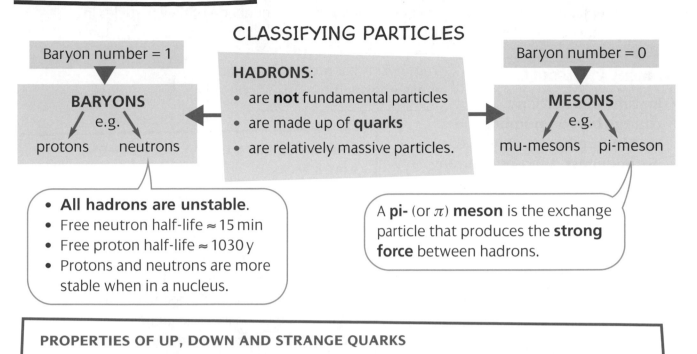

Baryon number = 1

**BARYONS**
e.g.
protons    neutrons

**HADRONS:**
* are **not** fundamental particles
* are made up of **quarks**
* are relatively massive particles.

Baryon number = 0

**MESONS**
e.g.
mu-mesons    pi-meson

* **All hadrons are unstable**.
* Free neutron half-life ≈ 15 min
* Free proton half-life ≈ 1030 y
* Protons and neutrons are more stable when in a nucleus.

A **pi- (or π) meson** is the exchange particle that produces the **strong force** between hadrons.

---

**PROPERTIES OF UP, DOWN AND STRANGE QUARKS**

* Evidence suggests that quarks do not exist in the free state.
* Other quarks exist and are given the properties of charm, topness and bottomness.

|  | Up quark u | Down quark d | Strange quark s | Up antiquark ū | Down antiquark d̄ | Strange antiquark s̄ |
|---|---|---|---|---|---|---|
| **Charge** | $+\frac{2}{3}$ | $-\frac{1}{3}$ | $-\frac{1}{3}$ | $-\frac{2}{3}$ | $+\frac{1}{3}$ | $+\frac{1}{3}$ |
| **Baryon number** | $+\frac{1}{3}$ | $+\frac{1}{3}$ | $+\frac{1}{3}$ | $-\frac{1}{3}$ | $-\frac{1}{3}$ | $-\frac{1}{3}$ |
| **Strangeness** | 0 | 0 | −1 | 0 | 0 | +1 |

* **Baryons** are made of **three** quarks.
* A **neutron** is one up quark and two down quarks (**udd**).
  Charge $= +\frac{2}{3} - \frac{1}{3} - \frac{1}{3} = 0$ Baryon number $= +\frac{1}{3} + \frac{1}{3} + \frac{1}{3} = 1$
* A **proton** is two up quarks and one down quark (**uud**).
  Charge $= +\frac{2}{3} + \frac{2}{3} - \frac{1}{3} = 1$ Baryon number $= +\frac{1}{3} + \frac{1}{3} + \frac{1}{3} = 1$

* **Mesons** are made of **two** quarks.
* A $\pi^+$ (**positive pion**) is one up and one antidown (ud̄).
  Charge $= +\frac{2}{3} + \frac{1}{3} = 1$
  Baryon number $= +\frac{1}{3} - \frac{1}{3} = 0$

---

**LEPTONS**

* Examples of leptons are electrons, muons and tau particles and their associated neutrinos.
* **Particles have lepton number 1:**.
  $e^-$ ($\beta^-$) $\mu^-$ $\tau^-$ and their associated neutrinos $\nu_e$ $\nu_\mu$ $\nu_\tau$
* **Antiparticles have lepton number −1:**
  $e^+$ ($\beta^+$) $\mu^+$ $\tau^+$ and the antineutrinos $\bar{\nu}_e$ $\bar{\nu}_\mu$ $\bar{\nu}_\tau$

Electrons are stable.

Mu and tau particles decay. (A tau is massive enough to decay into hadrons.)

**MUST REMEMBER**

When an electron is emitted in beta decay the other particle emitted is an **anti**neutrino.

Leptons are:
* thought to be fundamental particles
* much lighter than hadrons.

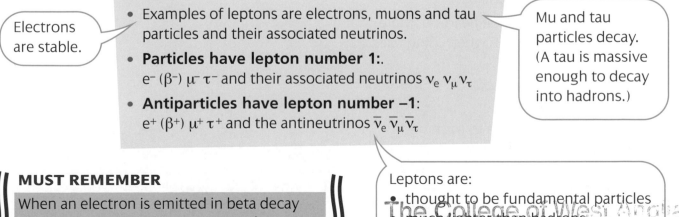

# PARTICLE INTERACTIONS

- Particle interactions include the decay of particles and the production of new particles in collisions between accelerated particles.

## CONSERVATION RULES

- In particle interactions:
**charge**, **baryon number** and **lepton number** are conserved.
(**Strangeness** is conserved in any interactions involving the strong nuclear force.)

> Remember that **momentum** and **mass-energy** must also be conserved.

### WORKED EXAMPLE

Do the conservation rules suggest that the following interaction is possible?

$$\overline{\nu}_\mu + p \rightarrow n + \mu^+$$

| | |
|---|---|
| Charge: | $0 + 1 \Rightarrow 0 + 1$  OK |
| Baryon number: | $0 + 1 \Rightarrow 1 + 0$  OK |
| Lepton number: | $-1 + 0 \Rightarrow 0 - 1$  OK |

Possible on basis of these rules.

- Neutron mass > proton mass
- Muon mass $\approx 200 \times$ mass of electron
- Neutrino has negligible mass (if any).
- Energy of the colliding particles must provide the extra mass.

Interaction conserves charge, baryon and lepton number.
Mass of protons $\approx 1880$ MeV
Mass of pions $\approx 3 \times 140 \approx 420$ MeV
$1880 - 420 = 1460$ MeV becomes kinetic energy of the pions

## EXPLAINING BETA DECAY USING THE QUARK MODEL

- This is an example of a weak interaction between the quarks.
- A neutron emits a proton and an intermediate W⁻ boson, which then decays into a beta⁻ particle and an antineutrino.
- In terms of quarks, udd becomes uud. A down quark changes to an up quark.
  $d \rightarrow u + W^-$ (intermediate boson)
  $\quad\quad \hookrightarrow \beta^- + \overline{\nu}$

## PARTICLE–ANTIPARTICLE ANNIHILATION

- When a particle and antiparticle collide they **annihilate** (destroy) each other.
- A slow-moving **electron and positron** produce two gamma ray photons.
  $$e^- + e^+ \rightarrow \gamma + \gamma$$
- Two $\gamma$-ray photons are produced so that momentum can be conserved.
- A **proton and antiproton** collision can produce the three types of pion: $\pi^+ \, \pi^- \, \pi^0$.
  $$p + \overline{p} \rightarrow \pi^- + \pi^+ + \pi^0$$

## HOW MASSIVE PARTICLES ARE MADE

- New particles are produced when accelerated particles collide.
- When two beams collide head-on, more energy is available than when using fixed targets.
- Electron–positron and proton–antiproton collisions are used.
- The total mass available to form new particles is the rest mass of the colliding particles plus their kinetic energy.

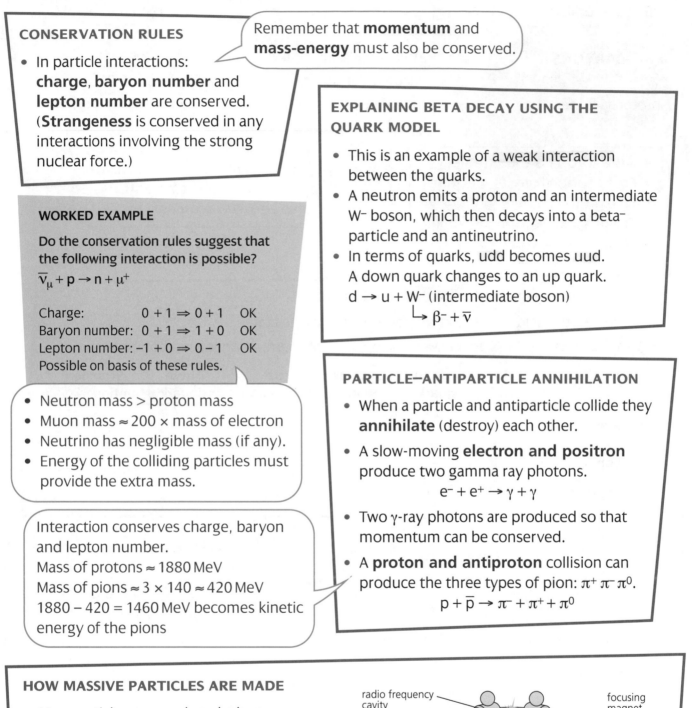

- When protons and antiprotons collide, they split up into quarks and antiquarks, which can then rearrange to produce new particles.
- The new particles can then decay to produce even more new particles.

# SECOND LAW OF THERMODYNAMICS

It is impossible to convert all the thermal energy in a source completely into work.

← Two forms of the **second law of thermodynamics** →

To convert energy from a hot source into work, a heat sink is necessary.

## ENERGY TRANSFER IN AN ENGINE

**Engine efficiency** = $\dfrac{\text{useful work done}}{\text{energy input}}$

**Efficiency** = $\dfrac{W}{Q_H}$

Using energy conservation $W = Q_H - Q_C$

**Efficiency** = $\dfrac{Q_H - Q_C}{Q_H}$

or

**Efficiency** = $1 - \dfrac{Q_C}{Q_H}$

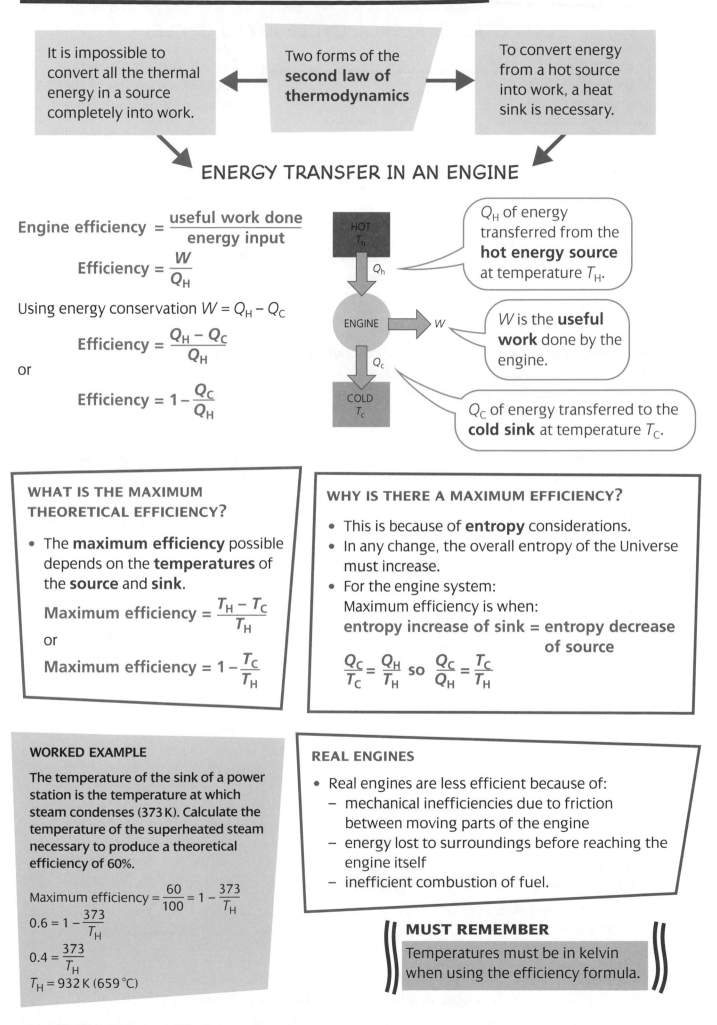

HOT $T_h$

$Q_h$

ENGINE → $W$

$Q_c$

COLD $T_c$

$Q_H$ of energy transferred from the **hot energy source** at temperature $T_H$.

$W$ is the **useful work** done by the engine.

$Q_C$ of energy transferred to the **cold sink** at temperature $T_C$.

---

### WHAT IS THE MAXIMUM THEORETICAL EFFICIENCY?

- The **maximum efficiency** possible depends on the **temperatures** of the **source** and **sink**.

  **Maximum efficiency** = $\dfrac{T_H - T_C}{T_H}$

  or

  **Maximum efficiency** = $1 - \dfrac{T_C}{T_H}$

### WHY IS THERE A MAXIMUM EFFICIENCY?

- This is because of **entropy** considerations.
- In any change, the overall entropy of the Universe must increase.
- For the engine system:
  Maximum efficiency is when:
  **entropy increase of sink = entropy decrease of source**

  $\dfrac{Q_C}{T_C} = \dfrac{Q_H}{T_H}$ so $\dfrac{Q_C}{Q_H} = \dfrac{T_C}{T_H}$

---

### WORKED EXAMPLE

The temperature of the sink of a power station is the temperature at which steam condenses (373 K). Calculate the temperature of the superheated steam necessary to produce a theoretical efficiency of 60%.

Maximum efficiency $= \dfrac{60}{100} = 1 - \dfrac{373}{T_H}$

$0.6 = 1 - \dfrac{373}{T_H}$

$0.4 = \dfrac{373}{T_H}$

$T_H = 932\,\text{K} \ (659\,^\circ\text{C})$

### REAL ENGINES

- Real engines are less efficient because of:
  - mechanical inefficiencies due to friction between moving parts of the engine
  - energy lost to surroundings before reaching the engine itself
  - inefficient combustion of fuel.

### MUST REMEMBER

Temperatures must be in kelvin when using the efficiency formula.

# ENGINE CYCLES

## FOUR-STROKE PETROL CYCLE (OTTO CYCLE)

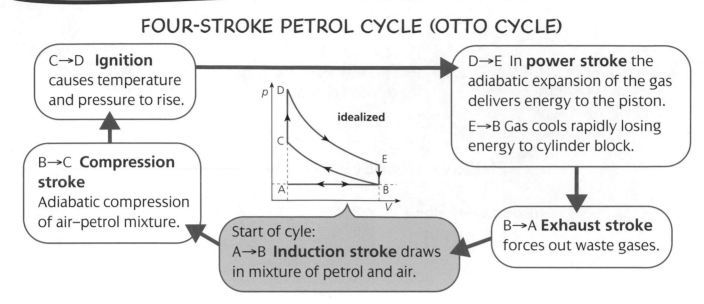

**C→D Ignition** causes temperature and pressure to rise.

**B→C Compression stroke** Adiabatic compression of air–petrol mixture.

**D→E In power stroke** the adiabatic expansion of the gas delivers energy to the piston.

**E→B** Gas cools rapidly losing energy to cylinder block.

**B→A Exhaust stroke** forces out waste gases.

Start of cyle: **A→B Induction stroke** draws in mixture of petrol and air.

idealized

- The **indicated power** of an engine is:

**area enclosed by the p-V loop** × **number of cycles per second** × **number of cylinders in the engine**

Area enclosed BCDE is the work done **in one cycle** (see page 50)

## DIESEL CYCLE

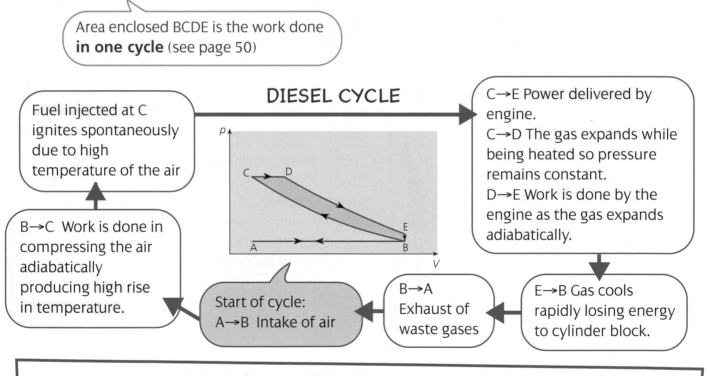

Fuel injected at C ignites spontaneously due to high temperature of the air

**B→C** Work is done in compressing the air adiabatically producing high rise in temperature.

**C→E** Power delivered by engine.
**C→D** The gas expands while being heated so pressure remains constant.
**D→E** Work is done by the engine as the gas expands adiabatically.

**E→B** Gas cools rapidly losing energy to cylinder block.

**B→A** Exhaust of waste gases

Start of cycle: **A→B** Intake of air

## HOW DO REAL ENGINES COMPARE WITH IDEAL CYCLES?

- Corners of the cycles are rounded because of time taken to open and close valves.
- There is not a fixed mass of gas in the cylinder during the cycle.
- Heat transfer does not take place at constant volume (**Otto cycle**) or constant pressure (**diesel cycle**).
- Pressure and temperature at different parts of the cylinder varies at any time.
- Heat transfer to cylinder occurs during compression and expansion so changes are not truly adiabatic.

**Overall effect of differences is a lower efficiency for real engines.**

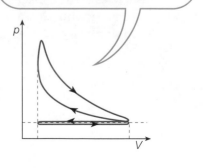

This is a four-stroke (Otto) cycle for a real engine.

# ROTATIONAL DYNAMICS

- Equations for **rotational motion** have the same form as those for **linear motion**.

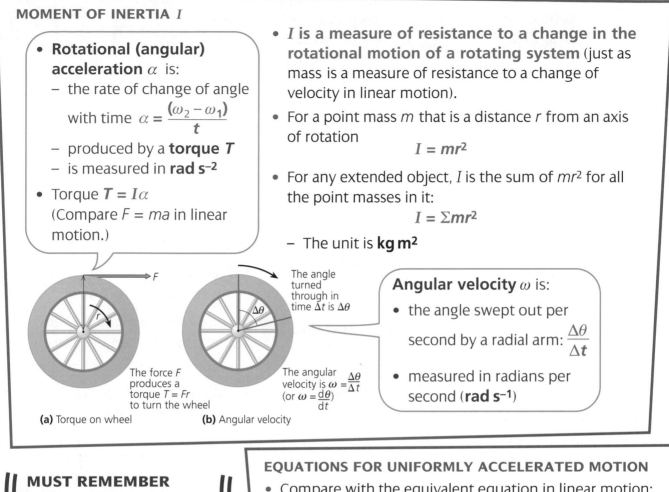

## MOMENT OF INERTIA $I$

- **Rotational (angular) acceleration** $\alpha$ is:
  - the rate of change of angle with time $\alpha = \dfrac{(\omega_2 - \omega_1)}{t}$
  - produced by a **torque $T$**
  - is measured in **rad s$^{-2}$**
- Torque $T = I\alpha$
  (Compare $F = ma$ in linear motion.)

- $I$ is a measure of resistance to a change in the rotational motion of a rotating system (just as mass is a measure of resistance to a change of velocity in linear motion).
- For a point mass $m$ that is a distance $r$ from an axis of rotation
  $$I = mr^2$$
- For any extended object, $I$ is the sum of $mr^2$ for all the point masses in it:
  $$I = \Sigma mr^2$$
  - The unit is **kg m$^2$**

$F$

The angle turned through in time $\Delta t$ is $\Delta\theta$

$\Delta\theta$

$r$

The force $F$ produces a torque $T = Fr$ to turn the wheel

**(a)** Torque on wheel

The angular velocity is $\omega = \dfrac{\Delta\theta}{\Delta t}$ (or $\omega = \dfrac{d\theta}{dt}$)

**(b)** Angular velocity

**Angular velocity $\omega$ is:**
- the angle swept out per second by a radial arm: $\dfrac{\Delta\theta}{\Delta t}$
- measured in radians per second (**rad s$^{-1}$**)

## MUST REMEMBER

$\omega_1$ = initial angular velocity
$\omega_2$ = final angular velocity
$\theta$ = angle turned through in time $t$
$\alpha$ = angular acceleration

## EQUATIONS FOR UNIFORMLY ACCELERATED MOTION

- Compare with the equivalent equation in linear motion:

| **Rotational motion** | **Linear motion** |
|---|---|
| $\omega_2 = \omega_1 + \alpha t$ | $v = u + at$ |
| $\theta = \omega_1 t + \frac{1}{2}\alpha t^2$ | $s = ut + \frac{1}{2}at^2$ |
| $\omega_2{}^2 = \omega_1{}^2 + 2\alpha\theta$ | $v^2 = u^2 + 2as$ |
| $\theta = \frac{1}{2}(\omega_1 + \omega_2)t$ | $s = \frac{1}{2}(u + v)t$ |

## WORKED EXAMPLE

The moment inertia of a wheel is 15 kg m$^2$. It is spinning at 30 revolutions per minute. A constant frictional force of 4.5 N acts at a distance of 0.15 m from the axis of rotation of a wheel. Calculate:
(a) the angular deceleration of the wheel
(b) the time taken to come to rest
(c) the number of revolutions before coming to rest.

(a) Torque $= 4.5 \times 0.15 = 0.675\,\text{N m}$
$$T = I\alpha$$
$$-0.675 = 15 \times \alpha$$
$$\alpha = -0.045\,\text{rad s}^{-2}$$
Negative acceleration means deceleration.

(b) Initial angular velocity $= \dfrac{2\pi \times 30}{60}\,\text{rad s}^{-1}$
$$= 3.14\,\text{rad s}^{-1}$$
$$\omega_2 = \omega_1 + \alpha t$$
$$0 = 3.14 - 0.045t$$
$$t = 70\,\text{s}$$

## MUST TAKE CARE

- Angular speeds are often given in revolutions per second or revolutions per minute. Convert these to rad s$^{-1}$.
- Remember to use **negative sign** for $\alpha$ when there is deceleration.

(c) $\theta = \omega_1 t + \frac{1}{2}\alpha t^2$
$$\theta = 3.14 \times 70 - \frac{1}{2}0.045 \times 70^2$$
$$\theta = 220 - 110 = 110\,\text{radians}$$
$$\theta = \frac{110}{2\pi} = 17.5\,\text{revolutions}$$

Any of the equations containing $\theta$ could be used here.

# CALCULATING WORK, ENERGY AND POWER IN ROTATING SYSTEMS

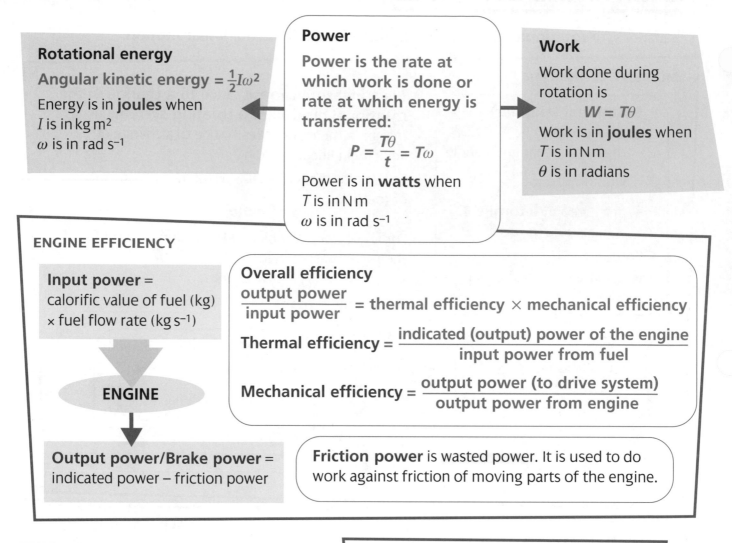

## Rotational energy

Angular kinetic energy $= \frac{1}{2}I\omega^2$

Energy is in **joules** when
$I$ is in kg m$^2$
$\omega$ is in rad s$^{-1}$

## Power

Power is the rate at which work is done or rate at which energy is transferred:

$$P = \frac{T\theta}{t} = T\omega$$

Power is in **watts** when
$T$ is in N m
$\omega$ is in rad s$^{-1}$

## Work

Work done during rotation is
$$W = T\theta$$
Work is in **joules** when
$T$ is in N m
$\theta$ is in radians

## ENGINE EFFICIENCY

**Input power =**
calorific value of fuel (kg)
× fuel flow rate (kg s$^{-1}$)

**ENGINE**

**Output power/Brake power =**
indicated power – friction power

### Overall efficiency

$$\frac{\text{output power}}{\text{input power}} = \text{thermal efficiency} \times \text{mechanical efficiency}$$

$$\text{Thermal efficiency} = \frac{\text{indicated (output) power of the engine}}{\text{input power from fuel}}$$

$$\text{Mechanical efficiency} = \frac{\text{output power (to drive system)}}{\text{output power from engine}}$$

**Friction power** is wasted power. It is used to do work against friction of moving parts of the engine.

## ANGULAR MOMENTUM

- **Angular momentum** $= I\omega$
  (Compare with $mv$ in linear motion.)
- Angular momentum can be changed by applying a **torque** $T$ for time $t$.
  This provides an **angular impulse** $= Tt$
  which changes the angular momentum.
  $$Tt = I(\omega_2 - \omega_1)$$

## CONSERVATION OF ANGULAR MOMENTUM

- Provided no external torques act on a system, the total angular momentum in any direction remains constant.

**WORKED EXAMPLE**

A drive is applied to a wheel of moment of inertia 4 kg m$^2$ that is spinning at 80 rad s$^{-1}$. It applies a torque of 24 N m for 4.5 s. Calculate the final angular velocity of the wheel.

Change in angular momentum $= Tt$
$$= 24 \times 4.5$$
$$= 108 \text{ N m s}$$

Change in angular velocity $= \frac{\Delta(I\omega)}{I} = \frac{108}{4}$
$$= 27 \text{ rad s}^{-1}$$
Final angular velocity $= 80 + 27 = 107 \text{ rad s}^{-1}$

**WORKED EXAMPLE**

A turntable spinning at 20 rad s$^{-1}$ has a moment of inertia of 1.2 kg m$^2$. A uniform disc of moment of inertia 0.60 kg m$^2$, spinning at 4.0 rad s$^{-1}$ in the opposite direction, is dropped on to it and sticks to it. The disc is dropped so that its mass is uniformly distributed about the axis of rotation. Calculate the final angular velocity of the turntable.

Final moment of inertia $= 1.2 + 0.6 = 1.8 \text{ kg m}^2$
Final angular momentum = initial angular momentum
$$1.8\omega = 1.2 \times 20 - 0.60 \times 4.0$$
Final angular velocity $= \frac{21.6}{1.8} = 12 \text{ rad s}^{-1}$

Rotation of turntable is still in original direction.

# LENSES

**Sign of *f***

**Positive** for a **converging** lens

**Negative** for a **diverging** lens

**Lens formula**

$$\frac{1}{f} = \frac{1}{u} + \frac{1}{v}$$

*u* is the object distance from the lens
*v* is the image distance
*f* is the focal length of the lens

**Signs of *u* and *v***

**Real** distances are **positive**.

**Virtual** distances are **negative**.

For a combination of lenses, the power is the sum of the individual powers.
Note that power is negative for a diverging lens.

## POWER OF A LENS

$$\text{Power} = \frac{1}{f}$$

$$\text{Power in dioptres (D)} = \frac{1}{\text{focal length in metres}}$$

**MAGNIFICATION PRODUCED BY A LENS**

$$M = \frac{\text{image size}}{\text{object size}} = \frac{\text{image distance}}{\text{object distance}} = \frac{v}{u}$$

### PROBLEMS WITH LENSES

- With **chromatic aberration**, the image is blurred because:
  - the focal length of the lens is different for different wavelengths of light
  - images for different wavelengths are at different distances from the lens.
- Chromatic aberration can be reduced using combinations of lenses.
- With **spherical aberration**, the image is blurred because:
  - rays entering the lens further from the axis deviate more than others
  - focal length is shorter for rays entering the lens further from the axis.
- Spherical aberration can be reduced by using a smaller aperture.

## CONVERGING LENS

**USE TO PRODUCE A REAL IMAGE**

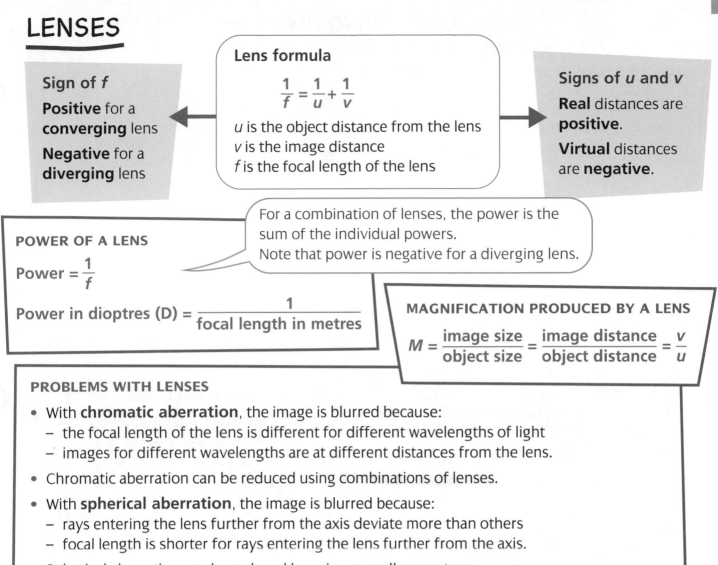

Ray parallel to the axis goes through the **principal focus**.

**Real image** can be focused on a screen. Rays of light really pass through the image.

Ray through the **principal focus** emerges parallel to the axis.

Path of ray through the **optical centre** is unchanged.

This image is **inverted** (upside-down when compared with the object).

### WORKED EXAMPLE

An object is 0.15 m from a lens that has a power of 10 D. Calculate the position of the image and magnification produced by the lens.

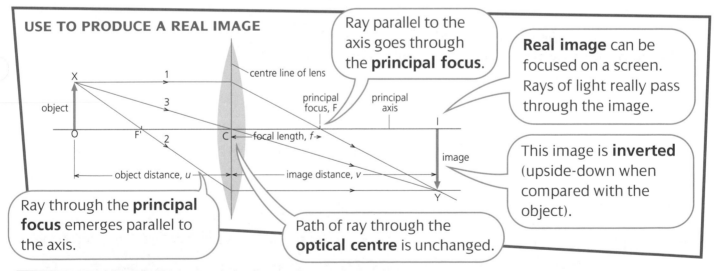

Focal length of lens $= \frac{1}{10} = 0.10\,\text{m}$

$\frac{1}{f} = \frac{1}{u} + \frac{1}{v} \Rightarrow \frac{1}{10} = \frac{1}{0.15} + \frac{1}{v}$

$\frac{1}{v} = \frac{1}{0.10} - \frac{1}{0.15} = 10.0 - 6.7 = 3.3\,\text{m}^{-1}$

$v = 0.30\,\text{m}$

Image is real and 0.30 m from the lens on the opposite side to the object.

Magnification $= \frac{v}{u} = \frac{0.30}{0.15} = 2$

Image is twice the size of the object.

# CONVERGING LENS

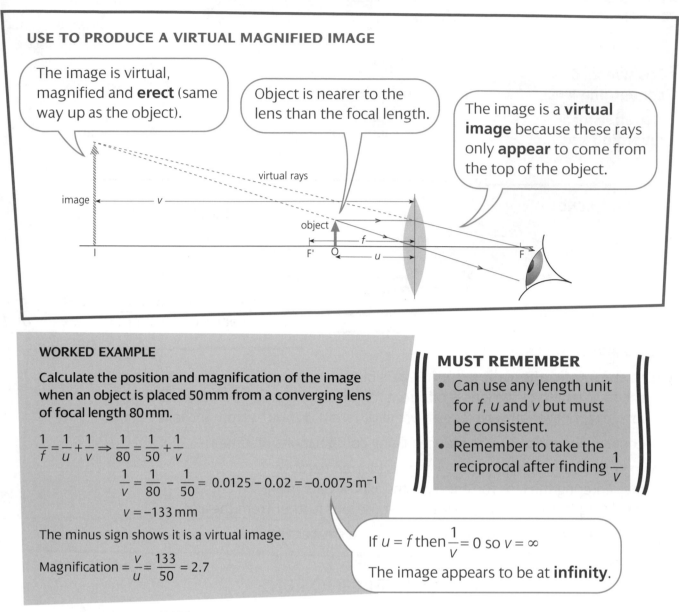

## USE TO PRODUCE A VIRTUAL MAGNIFIED IMAGE

The image is virtual, magnified and **erect** (same way up as the object).

Object is nearer to the lens than the focal length.

The image is a **virtual image** because these rays only **appear** to come from the top of the object.

virtual rays

image — $v$ — 

object

$f$

I    F'    O    — $u$ —    F

---

**WORKED EXAMPLE**

Calculate the position and magnification of the image when an object is placed 50 mm from a converging lens of focal length 80 mm.

$$\frac{1}{f} = \frac{1}{u} + \frac{1}{v} \Rightarrow \frac{1}{80} = \frac{1}{50} + \frac{1}{v}$$

$$\frac{1}{v} = \frac{1}{80} - \frac{1}{50} = 0.0125 - 0.02 = -0.0075 \text{ m}^{-1}$$

$$v = -133 \text{ mm}$$

The minus sign shows it is a virtual image.

$$\text{Magnification} = \frac{v}{u} = \frac{133}{50} = 2.7$$

**MUST REMEMBER**

- Can use any length unit for $f$, $u$ and $v$ but must be consistent.
- Remember to take the reciprocal after finding $\frac{1}{v}$

If $u = f$ then $\frac{1}{v} = 0$ so $v = \infty$

The image appears to be at **infinity**.

---

# DIVERGING LENS

- A **diverging lens** always produces a **virtual image** of a real object.

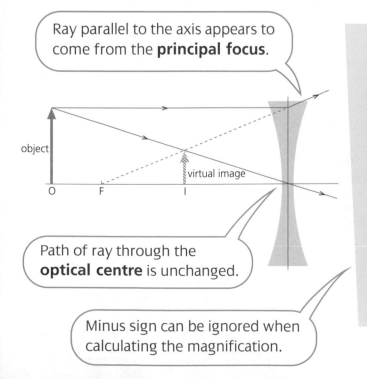

Ray parallel to the axis appears to come from the **principal focus**.

object

O    F    I

virtual image

Path of ray through the **optical centre** is unchanged.

Minus sign can be ignored when calculating the magnification.

**WORKED EXAMPLE**

Calculate the position and magnification of the image when an object is placed 20 cm from a lens of focal length 15 cm.

$$\frac{1}{f} = \frac{1}{u} + \frac{1}{v} \Rightarrow -\frac{1}{15} = \frac{1}{20} + \frac{1}{v}$$

$$\frac{1}{v} = -\frac{1}{15} - \frac{1}{20} = -0.067 - 0.050 = -0.117 \text{ cm}^{-1}$$

$$v = -8.5 \text{ cm}$$

The image is virtual. It will be on the same side of the lens as the object.

$$\text{Magnification} = \frac{v}{u} = \frac{8.5}{20} = 0.43$$

The image is diminished (0.43 times the size of the object).

# THE EYE

## STRUCTURE OF THE EYE

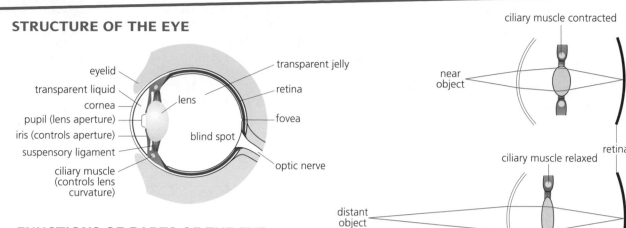

## FUNCTIONS OF PARTS OF THE EYE

- **Cornea** is a fixed focus lens and does most of focusing of light.

- **Lens** has a variable focus and is used to finely adjust the focusing of the image.

- **Ciliary muscles** adjusts the curvature (focal length) of the lens.

- **Pupil** is the region where the light enters the eye.

- **Iris** controls the amount of light entering the eye by controlling the diameter of the pupil.

- **Retina** consists of light-sensitive cells (**rods** and **cones**), which respond to the light that falls on them.

- **Optic nerve** takes signals generated by the rods and cones to the brain.

- **Vitreous humour** fills the eye so that it keeps its shape.

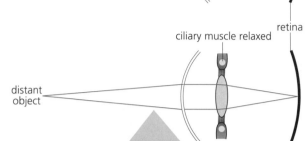

### Accommodation

- This is the ability to focus objects at different distances from the eye.

- The lens is made fatter and more curved to focus nearer objects.

- When the eye is looking at very distant objects, ciliary muscles are fully relaxed.

- **Near point** is the closest point an object can be to the eye and be clearly focused (about 250 mm for a normal eye).

- **Far point** is the furthest point an object can be from the eye and be clearly focused (infinity for a normal eye).

- **Depth of field** is the distance between the near point and the far point.

## FUNCTION OF RODS AND CONES

- **Rods** are not colour sensitive. They respond to light intensity and are particularly sensitive in low light conditions (**scotopic vision**).

- **Cones** are colour sensors. They need high light levels to function. The signals from the three types of cone are interpreted by the brain and enable the full range of colour to be seen (**photopic vision**).

- **Colour blindness** results when one or more types of cone are missing or have reduced sensitivity.

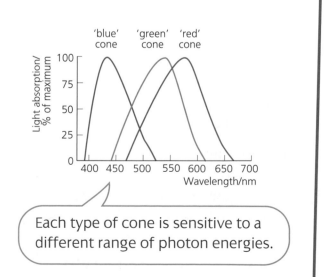

Each type of cone is sensitive to a different range of photon energies.

# DEFECTS OF VISION

## SHORT SIGHT (MYOPIA)

- The eye can see objects close to it, but not distant objects.
- The focal length of the lens is too small when the ciliary muscles are relaxed.
- The far point is not at infinity.

> Distant objects are focused in **front of the retina**.

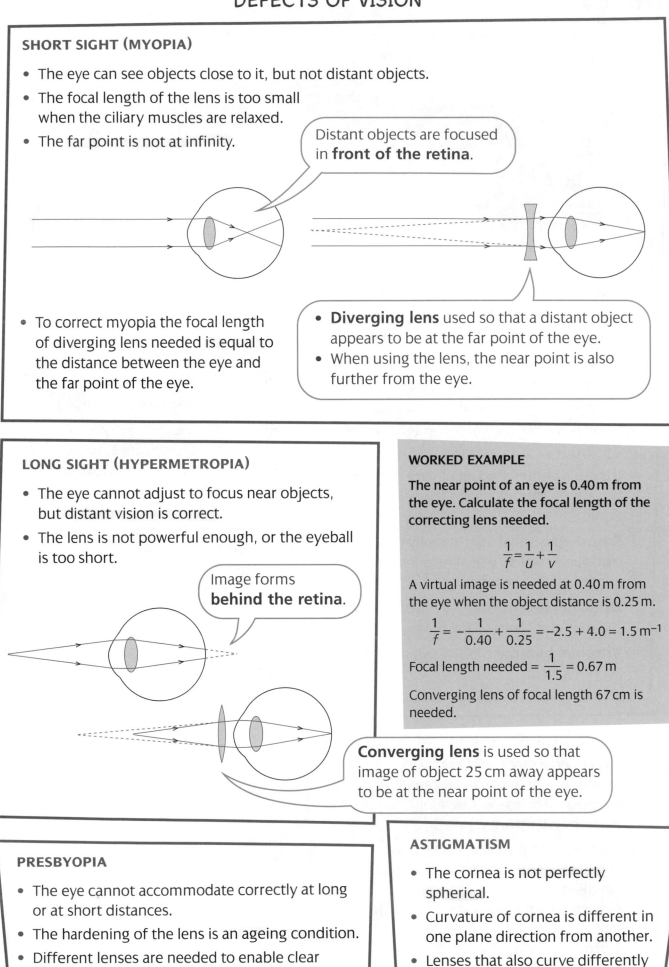

- To correct myopia the focal length of diverging lens needed is equal to the distance between the eye and the far point of the eye.

- **Diverging lens** used so that a distant object appears to be at the far point of the eye.
- When using the lens, the near point is also further from the eye.

## LONG SIGHT (HYPERMETROPIA)

- The eye cannot adjust to focus near objects, but distant vision is correct.
- The lens is not powerful enough, or the eyeball is too short.

> Image forms **behind the retina**.

**Converging lens** is used so that image of object 25 cm away appears to be at the near point of the eye.

### WORKED EXAMPLE

The near point of an eye is 0.40 m from the eye. Calculate the focal length of the correcting lens needed.

$$\frac{1}{f} = \frac{1}{u} + \frac{1}{v}$$

A virtual image is needed at 0.40 m from the eye when the object distance is 0.25 m.

$$\frac{1}{f} = -\frac{1}{0.40} + \frac{1}{0.25} = -2.5 + 4.0 = 1.5\,\text{m}^{-1}$$

Focal length needed $= \dfrac{1}{1.5} = 0.67\,\text{m}$

Converging lens of focal length 67 cm is needed.

## PRESBYOPIA

- The eye cannot accommodate correctly at long or at short distances.
- The hardening of the lens is an ageing condition.
- Different lenses are needed to enable clear focusing of near and distant objects.
- Bifocal or varifocal lenses are used.

## ASTIGMATISM

- The cornea is not perfectly spherical.
- Curvature of cornea is different in one plane direction from another.
- Lenses that also curve differently in one plane from another are used to compensate.

# THE EAR

- An ear is a **transducer** that converts sound waves into electrical impulses.

Sound waves cause the **drum skin** to vibrate with the same frequency as the wave.

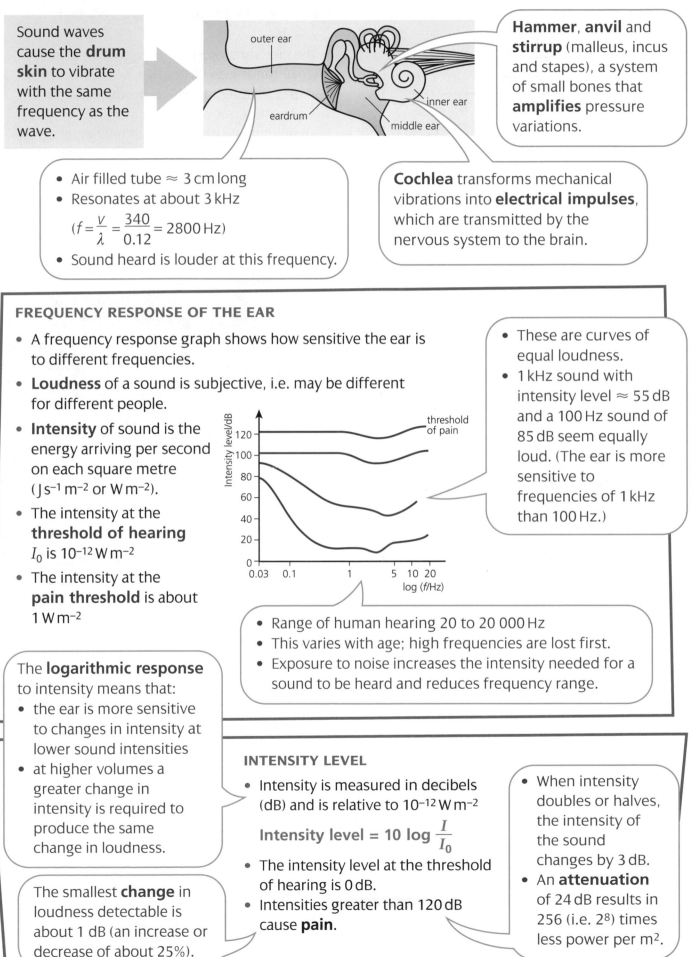

**Hammer**, **anvil** and **stirrup** (malleus, incus and stapes), a system of small bones that **amplifies** pressure variations.

- Air filled tube $\approx 3\,\text{cm}$ long
- Resonates at about $3\,\text{kHz}$
  $(f = \dfrac{v}{\lambda} = \dfrac{340}{0.12} = 2800\,\text{Hz})$
- Sound heard is louder at this frequency.

**Cochlea** transforms mechanical vibrations into **electrical impulses**, which are transmitted by the nervous system to the brain.

## FREQUENCY RESPONSE OF THE EAR

- A frequency response graph shows how sensitive the ear is to different frequencies.
- **Loudness** of a sound is subjective, i.e. may be different for different people.
- **Intensity** of sound is the energy arriving per second on each square metre ($Js^{-1}m^{-2}$ or $Wm^{-2}$).
- The intensity at the **threshold of hearing** $I_0$ is $10^{-12}\,Wm^{-2}$
- The intensity at the **pain threshold** is about $1\,Wm^{-2}$

- These are curves of equal loudness.
- $1\,\text{kHz}$ sound with intensity level $\approx 55\,\text{dB}$ and a $100\,\text{Hz}$ sound of $85\,\text{dB}$ seem equally loud. (The ear is more sensitive to frequencies of $1\,\text{kHz}$ than $100\,\text{Hz}$.)

- Range of human hearing 20 to 20 000 Hz
- This varies with age; high frequencies are lost first.
- Exposure to noise increases the intensity needed for a sound to be heard and reduces frequency range.

The **logarithmic response** to intensity means that:
- the ear is more sensitive to changes in intensity at lower sound intensities
- at higher volumes a greater change in intensity is required to produce the same change in loudness.

The smallest **change** in loudness detectable is about 1 dB (an increase or decrease of about 25%).

## INTENSITY LEVEL

- Intensity is measured in decibels (dB) and is relative to $10^{-12}\,Wm^{-2}$

  **Intensity level = $10 \log \dfrac{I}{I_0}$**

- The intensity level at the threshold of hearing is 0 dB.
- Intensities greater than 120 dB cause **pain**.

- When intensity doubles or halves, the intensity of the sound changes by 3 dB.
- An **attenuation** of 24 dB results in 256 (i.e. $2^8$) times less power per m².

# X-RAYS

## PRODUCING X-RAYS

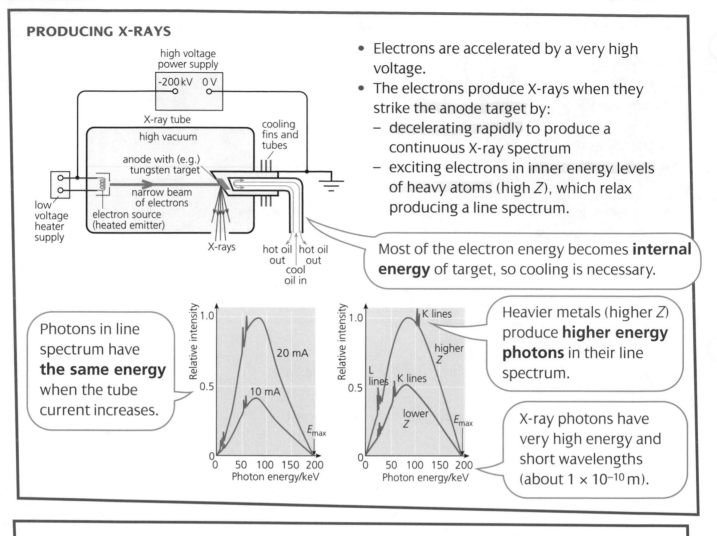

- Electrons are accelerated by a very high voltage.
- The electrons produce X-rays when they strike the anode target by:
  - decelerating rapidly to produce a continuous X-ray spectrum
  - exciting electrons in inner energy levels of heavy atoms (high $Z$), which relax producing a line spectrum.

Most of the electron energy becomes **internal energy** of target, so cooling is necessary.

Photons in line spectrum have **the same energy** when the tube current increases.

Heavier metals (higher $Z$) produce **higher energy photons** in their line spectrum.

X-ray photons have very high energy and short wavelengths (about $1 \times 10^{-10}$ m).

## X-RAYS IMAGING

### X-ray advantages
- Avoids invasive investigations (operations).
- Provides very quick diagnosis.

### X-ray disadvantages
- Risks of cell damage by ionisation.
- Poor distinction between soft tissues.

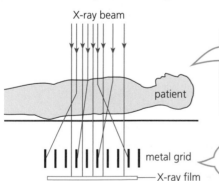

**Definition of image** can be improved by:
- a contrast medium (swallowed or injected), such as barium sulphate (barium meal)
- use of image intensifiers (phosphors in front of the film) that emit light onto a film after absorbing X-rays.

The metal grid removes scattered X-rays, improving contrast.

The transmitted intensity of an X-ray beam depends on the thickness $x$ of the absorber and the nature of the absorbing medium:

$$I = I_0 e^{-\mu x}$$

where $\mu$ is a constant that depends on the material. Dense material (bone) absorbs better than lighter material (soft tissue) elements.

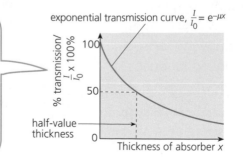

exponential transmission curve, $\frac{I}{I_0} = e^{-\mu x}$

# IMAGING WITH LIGHT

- An **endoscope**:
  - can be inserted into inaccessible places (e.g. into the digestive system, heart and lungs) to provide images to aid diagnosis and treatment
  - uses **total internal reflection** or **refraction** of light down bundles of optical fibres to illuminate and transmit images of the inside of the body.

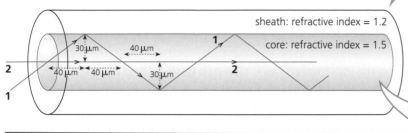

sheath: refractive index = 1.2
core: refractive index = 1.5

> There are many fibres like this each ≈ 10 μm diameter.

- Outer sheath has lower refractive index so **total internal reflection** occurs at the interface between the core and sheath.
- For figures given, the **critical angle c** is given by:
$$n_1 \sin c = n_2 \sin 90°$$
$$1.5 \sin c = 1.2$$
$$c = 53°$$

### WHAT PROBLEMS OCCUR WITH FIBRE OPTICS?

- Deterioration of image quality occurs due to:
  - different routes for light down the fibres (path 1 is longer than path 2) but this is improved using thinner fibres
  - misalignment of fibres
  - reflection of light at the ends of the tube
  - leakage of light from one fibre to another
  - attenuation of light as it passes down the fibre (about 50% per 2 metre length).

> The shaft is about 10 mm diameter and can contain controls for instruments that allow surgical operations (e.g. to treat tumours, ulcers, constrictions).

> The image can be viewed directly with a magnifying eyepiece or a camera can be fitted to produce an image on a screen.

> The fibres in the light tube can be in an **incoherent bundle**. It doesn't matter if they change positions along the tube.

> Using thinner fibres enables more fibres to be used to improve **resolution** (i.e. sharper images).

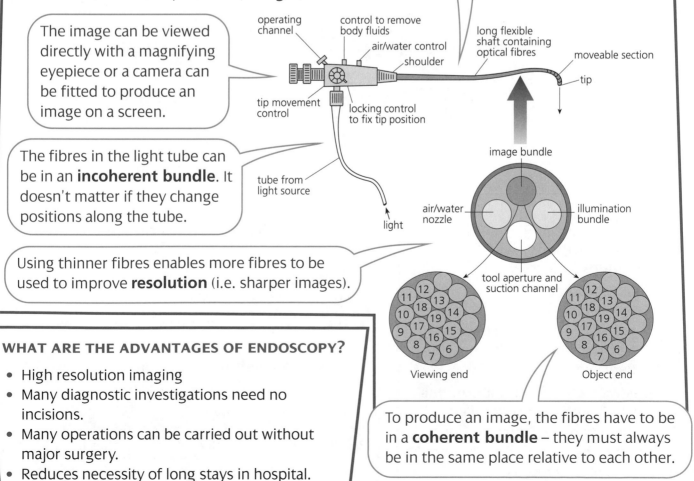

operating channel · control to remove body fluids · air/water control · shoulder · long flexible shaft containing optical fibres · moveable section · tip · tip movement control · locking control to fix tip position · tube from light source · light · image bundle · air/water nozzle · illumination bundle · tool aperture and suction channel · Viewing end · Object end

> To produce an image, the fibres have to be in a **coherent bundle** – they must always be in the same place relative to each other.

### WHAT ARE THE ADVANTAGES OF ENDOSCOPY?

- High resolution imaging
- Many diagnostic investigations need no incisions.
- Many operations can be carried out without major surgery.
- Reduces necessity of long stays in hospital.

# STELLAR EVOLUTION

## ENERGY PRODUCTION PROCESSES IN STARS

**FORMATION OF STARS**

- Hydrogen gas collapses due to gravitational attraction.
- P.E. is converted to internal energy of the gas so temperature rises.
- Nuclear fusion can take place when temperature reaches about $1 \times 10^8$ K.

**NUCLEAR FUSION PROCESSES IN THE SUN**

- Energy from the Sun is the result of two fusion processes – the **proton–proton cycle** and the **carbon cycle**.

Two protons produce a deuteron, positron and neutrino.

A positron **annihilates** with an electron producing gamma radiation.

A deuteron and a proton produce a helium-3 nucleus and gamma radiation.

Two helium-3 nuclei produce a helium-4 nucleus and two protons.

Overall four protons have combined to form a helium-4 nucleus and 28 MeV.

**Proton–proton chain**

$$^1_1p + {}^1_1p \Rightarrow {}^2_1H + {}^0_{+1}e + {}^0_0\nu$$
$$^0_{+1}e + {}^0_{-1}e \Rightarrow 2\,{}^0_0\gamma$$
$$^2_1H + {}^1_1p \Rightarrow {}^3_2He + {}^0_0\gamma$$
$$^3_2He + {}^3_2He \Rightarrow {}^4_2He + {}^1_1p + {}^1_1p$$

**Net reaction**

$$^1_1p + {}^1_1p + {}^1_1p + {}^1_1p + 2\,{}^0_{-1}e \Rightarrow {}^4_2He + \gamma \text{ and } \nu \text{ radiation}$$

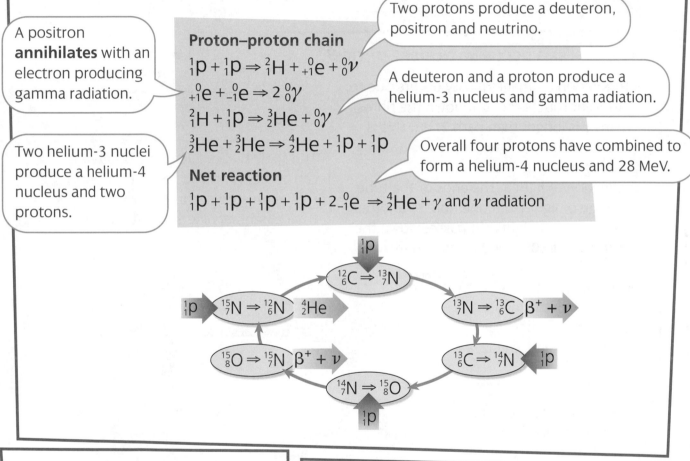

**WHAT HAPPENS NEXT?**

- Stars the size of the Sun form a **red giant**.
- Energy is produced by:
  - two helium-4 nuclei fusing to form beryllium-8, followed by
  - a helium-4 nucleus fusing with beryllium-8 to form carbon-12.
- Massive stars become **supernovae** in which lighter elements (like carbon) undergo further fusion to produce heavy elements (like iron) with explosive results.

**HOW LONG CAN THE SUN GO ON PRODUCING ENERGY?**

- The Sun contains about $1 \times 10^{29}$ kg of hydrogen, i.e. $10^{29} \times 10^3 \times 6 \times 10^{23} = 6 \times 10^{55}$ atoms
- The fusion of 4 atoms produces about 28 MeV or $4.5 \times 10^{-12}$ J of energy.
- Total energy available $= \frac{1}{4} \times 6 \times 10^{55} \times 4.5 \times 10^{-12}$ J
$$= 6.8 \times 10^{43} \text{ J}$$
- Sun emits $3.9 \times 10^{26}$ W
- Sun could emit energy for $\dfrac{6.8 \times 10^{43}}{3.9 \times 10^{26}} = 1.7 \times 10^{17}$ s
$$= 5.5 \times 10^9 \text{ y}$$

# HERSPRUNG-RUSSELL DIAGRAM

Scale uses luminosity instead of magnitude.

**Red giants** are massive stars with low surface temperature. (Low power emitted per m² but very bright due to large surface area.)

These are massive stars (≈30 × mass of Sun). Low lifetimes (≈1 × 10⁶ y) due to high density and temperature of core so uses hydrogen faster.

**Main sequence** stars include the Sun. Expected lifetime 1× 10¹⁰ y.

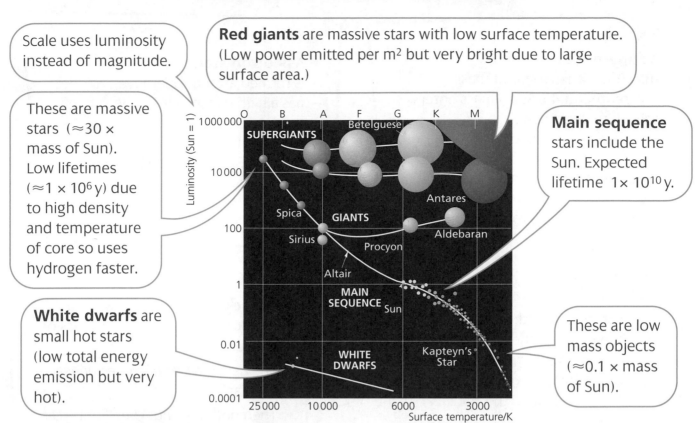

**White dwarfs** are small hot stars (low total energy emission but very hot).

These are low mass objects (≈0.1 × mass of Sun).

## SHOWING THE FATE OF A SUN-SIZED STAR

### Stage 3 Star becomes a white dwarf
- In a red giant, helium nuclei fuse to form beryllium.
- Beryllium combines with helium to form carbon.
- When this process is complete no further fusion is possible.
- Star becomes a **white dwarf**.
- Ultimately it cools to become an invisible **black dwarf**.

### Stage 2 Thermal pulsing occurs
- Energy output varies.
- Star goes through long periods of relatively low luminosity followed by explosive reactions due to release of energy by rapid fusion of helium nuclei.
- Star luminosity increases for short time then decreases again.

### Stage 1 Star becomes a red giant
- Fusion in core stops.
- Core cools down so pressure falls.
- Star collapses due to gravity.
- Temperature rises due to gravitational P.E. loss and energy from fusion in outer shell.
- Star expands since temperature rises as helium fuses to form heavier elements.
- Surface cools but has much larger radius so overall it becomes more luminous.

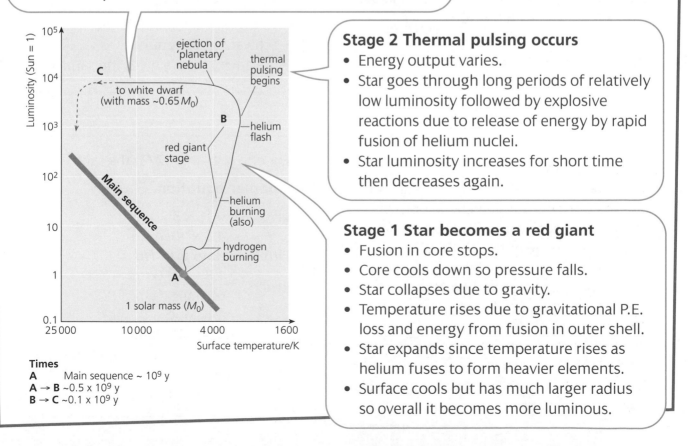

**Times**
A        Main sequence ~ 10⁹ y
A → B ~0.5 x 10⁹ y
B → C ~0.1 x 10⁹ y

# ASTRONOMICAL OBJECTS

## SUPERNOVAE

- A **supernova**:
- – may form when a star of mass greater than 1.4 × Sun's mass enters the white dwarf stage
- – is caused by a sudden collapse in which electrons and protons in the core combine to form neutrons
- – ejects heavy nuclei from outer layers of the supernova into space.

## NEUTRON STARS

- A **neutron star**:
- – is formed after a supernova explodes
- – has a high density ($10^{12}$ × density of the Earth)
- – has a radius ≈ 10 km
- – rotates rapidly (about 30 revolutions per second)
- – emits pulsing radio waves that are detected as a **pulsar**.

## What is the Schwarzschild radius?

- The **Schwarzschild radius**:
  - – is the radius of the **event horizon**
  - – **is the distance from the centre of a black hole** at which the escape velocity is equal to the velocity of light.
- Inside the event horizon, nothing is visible and nothing can escape.

## BLACK HOLES

- A **black hole** is formed when a neutron star exceeds 2.5 × mass of the Sun.
- The neutron star becomes denser as it collapses.
- Then nothing – not even light – can escape from it.
- It continues to attract matter so increases in size.

## HOW TO CALCULATE THE EVENT HORIZON

- The Schwarzschild radius for a mass $M$ is given by:

$$R_S = \frac{2GM}{c^2}$$

- For a mass the size of the Earth, $R_S$ would be:

$$\frac{2 \times 6.7 \times 10^{-11} \times 6.0 \times 10^{24}}{(3 \times 10^8)^2} \approx 9\,\text{mm}$$

- All the mass of the Earth would have to be confined in a sphere of diameter just 2 cm.

- Simple theory gives escape velocity:

$$v = \sqrt{\frac{2GM}{r^2}}$$

(See page 61.)
- When $v = c$ the radius is the Schwarzschild radius.
(Relativity theory gives the same result.)

## QUASARS

- **Quasars** (radio stars):
  - – are detected by the emission of radio waves
  - – are the most distant detectable objects
  - – move at speeds near to the speed of light so
    - ○ are at the edge of the observable Universe
    - ○ produce very large **red shifts**
  - – are the oldest detectable objects in the Universe
  - – emit as much energy as a whole galaxy the size of the Solar System
  - – are thought to be galaxies with massive black holes at their centre.

- For one quasar (3C273) the ratio $\frac{\Delta\lambda}{\lambda} = 0.16$
- **Doppler equation** $\frac{\Delta\lambda}{\lambda} = \frac{v}{c}$

gives $v = 0.16 \times 3 \times 10^8$
$= 4.8 \times 10^7 \,\text{m s}^{-1}$

**Hubble's law** $v = Hd$

gives $d = \dfrac{4.8 \times 10^4 \,(\text{km s}^{-1})}{65 \,(\text{km s}^{-1}\,\text{Mpc}^{-1})}$

$= 740\,\text{Mpc}$
$= 2.4 \times 10^9$ light years
(1 pc = 3.26 light years)

This suggests that quasars are being observed as they were $2.4 \times 10^9$ years ago.

# CLASSIFICATION OF STARS

## CLASSIFICATION OF STARS BY LUMINOSITY

### APPARENT MAGNITUDE $m$

- This relates to how bright the star appears to an observer.
- The scale is based on the dimmest and brightest objects visible with the naked eye.
- There are six levels of magnitude:
  - **first magnitude** $\Rightarrow$ brightest visible object
  - **sixth magnitude** $\Rightarrow$ just visible with naked eye.
- Magnitude is related to intensity $I$ detected by observer by:
$$m = -2.5 \log I + c$$
  where $c$ is a constant.
- An increase of 1 in magnitude = 2.51 decrease in intensity (power detected by observer per $m^2$).
- The sixth magnitude is $2.51^5$ (100) times less bright than the first magnitude.

> A dim star that is closer to Earth than a brighter star may appear equally bright.

### ABSOLUTE MAGNITUDE $M$

- This is used to compare the real brightness of objects.
- $M$ is the brightness an object would have if it were at a distance of 10 parsecs from the Earth.
- For a star at distance $r$, the difference between apparent and absolute brightness is:
$$m - M = 5 \log \left(\frac{r}{10}\right)$$
  where $r$ must be in parsecs (pc).
- 1 parsec (pc) = $3.09 \times 10^{16}$ m = 3.26 light years

### WHY IS $m - M = 5 \log \left(\frac{r}{10}\right)$?

**OCR only**

- Intensity at 10 pc = $I_{10} = \dfrac{P}{4\pi10^2}$

  Intensity at $r$ = observed intensity = $I = \dfrac{P}{4\pi r^2}$

$$\frac{I_{10}}{I} = \frac{r^2}{10^2}$$

$m = -2.5 \log I + c$ and $M = -2.5 \log I_{10} + c$

$m - M = -2.5 \log I - (-2.5 \log I_{10}) = 2.5 \log \left(\frac{I_{10}}{I}\right)$

$m - M = 2.5 \log \left(\frac{r^2}{10^2}\right) = 5 \log \left(\frac{r}{10}\right)$

### WORKED EXAMPLE

A star has an apparent brightness of 1.5 and is 60 pc from the Earth. Calculate the absolute brightness of the star.

$m - M = 5 \log \left(\frac{r}{10}\right) \Rightarrow 1.5 - M = 5 \log \left(\frac{60}{10}\right)$

$M = 1.5 - 5 \log 6 = 1.5 - 3.9 = -2.4$

## CLASSIFICATION OF STARS BY TEMPERATURE

- The colour of a star depends on the surface temperature.
- The wavelengths absorbed as light passes through the outer atmosphere of a star depends on its temperature.
- **Stefan's and Wien's laws** can be used to estimate the surface temperature of a star.
- The spectrum produced also provides evidence of the chemical composition of a star and hence its stage of evolution.

| Star type | Colour | Temperature/K | Chemical characteristics |
|---|---|---|---|
| O | Blue | 30 000 | Ionised helium (helium absorption spectrum) |
| B | Blue-white | 11 000–30 000 | Neutral helium and hydrogen (Balmer lines seen) |
| A | White | 7500–11 000 | Hydrogen and some ionised calcium |
| F | White to yellow | 6000–7500 | Ionised calcium and metals developing |
| G | Yellow | 5000–6000 | Ionised calcium lines, iron (little hydrogen) |
| K | Orange | 3500–5000 | Lot of neutral metal lines, some molecules (e.g. CH) |
| M | Red | About 3500 | Molecules such as titanium oxide (band spectra seen) |

> A red star having a **lower** temperature than a blue star may appear **brighter** because:
> - it is closer to the Earth
> - it has a much larger surface area emitting light.

# CONTINUOUS SPECTRA

- Hot bodies such as stars, filament lamps and molten steel give rise to **continuous spectra**.
- The radiation contains all the wavelengths that are possible at the temperature of the body.
- A **black body** is a perfect absorber of radiation (so it looks black).
- A **perfect absorber** is also a **perfect emitter**.
- Bodies that emit a continuous spectrum are also called black bodies.
- The spectrum is produced by many different kinds of energy transitions in a source.

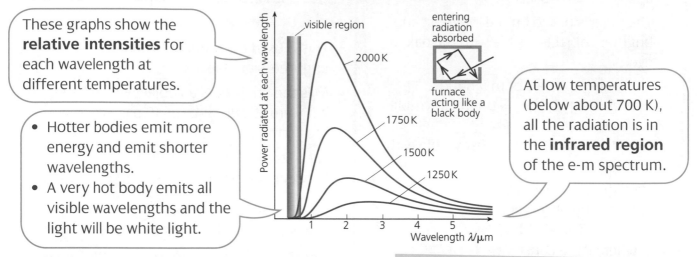

> These graphs show the **relative intensities** for each wavelength at different temperatures.

- Hotter bodies emit more energy and emit shorter wavelengths.
- A very hot body emits all visible wavelengths and the light will be white light.

> At low temperatures (below about 700 K), all the radiation is in the **infrared region** of the e-m spectrum.

- **Stefan's law** relates the total power $P$ radiated by a black body to its Kelvin temperature $T$:

$$P = \sigma A T^4$$

where $\sigma$ is **Stefan's constant** ($5.7 \times 10^{-8}$ W m$^{-2}$ K$^{-4}$) and $A$ is the area of the radiating surface.

- **Wien's displacement law** relates $\lambda_{max}$, the wavelength at which the peak power is radiated, to the kelvin temperature $T$:
$\lambda_{max} T$ = **constant** ($= 2.9 \times 10^{-3}$ m K)

- The temperature of a star can be found if the wavelength for peak power is measured.

**WORKED EXAMPLE**

The power output of the Sun is $3.9 \times 10^{26}$ W. Its surface area is $6.1 \times 10^{18}$ m$^2$ and its surface temperature is 6000 K. Calculate the surface area of a star that has the same power output but is at a surface temperature of 10 000 K.

For same power output, $AT^4$ = constant
$A \times 10\,000^4 = 6.1 \times 10^{18} \times 6000^4$
$A = \dfrac{6000^4}{10\,000^4} \times 6.1 \times 10^{18} = 0.60^4 \times 6.1 \times 10^{18}$
Surface area of star = $7.9 \times 10^{17}$ m$^2$

# OLBERS' PARADOX

**THE PARADOX**

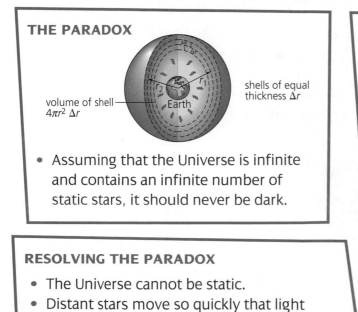

volume of shell
$4\pi r^2 \Delta r$

shells of equal thickness $\Delta r$

- Assuming that the Universe is infinite and contains an infinite number of static stars, it should never be dark.

**RESOLVING THE PARADOX**

- The Universe cannot be static.
- Distant stars move so quickly that light emitted from them never reaches the Earth.

**THE REASONING BEHIND THE PARADOX**

- There will be $N$ stars per unit volume. Average power emitted by a star = $P$
 Volume of a shell of thickness $\Delta r$ at a distance $r$ from Earth = $4\pi r^2 \Delta r$
 Number of stars in shell = $4\pi r^2 \Delta r N$
 Intensity of light from the shell = $\dfrac{4\pi r^2 NP \Delta r}{r^2}$
 $= 4\pi NP \Delta r$

- All shells produce the same intensity of light on the Earth. The sky should be equally bright in all directions so it should never be dark. An infinite number of shells should produce infinite brightness.

# TELESCOPES

- An **astronomical telescope** is used to view objects that are a long way off.
- The rays arriving are virtually parallel so that the object is effectively at infinity.
- In **normal adjustment** the telescope is adjusted to form the final image at infinity.

## REFRACTING TELESCOPES

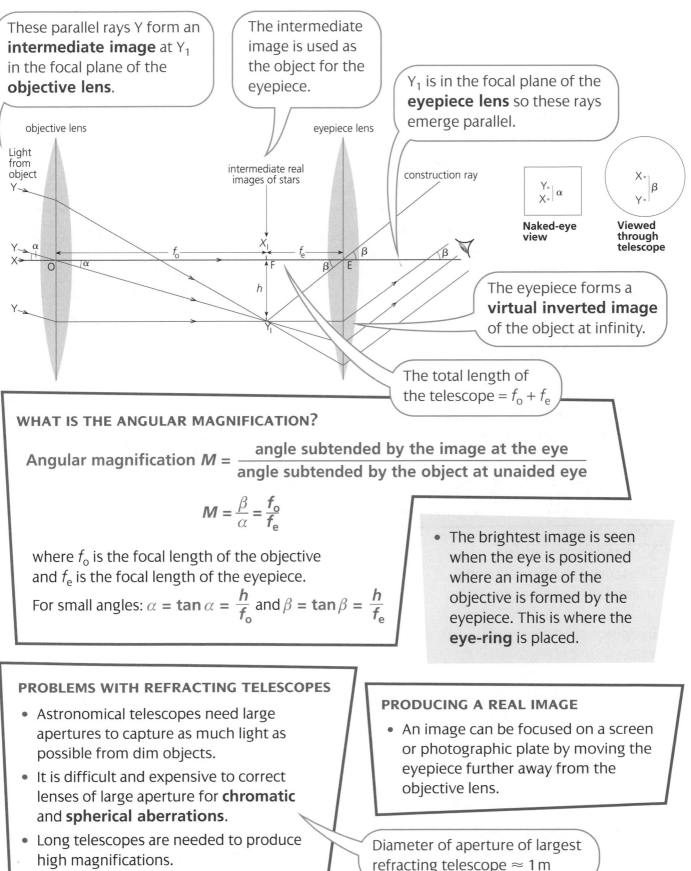

These parallel rays Y form an **intermediate image** at $Y_1$ in the focal plane of the **objective lens**.

The intermediate image is used as the object for the eyepiece.

$Y_1$ is in the focal plane of the **eyepiece lens** so these rays emerge parallel.

objective lens

eyepiece lens

Light from object Y

intermediate real images of stars

construction ray

Naked-eye view

Viewed through telescope

The eyepiece forms a **virtual inverted image** of the object at infinity.

The total length of the telescope $= f_o + f_e$

### WHAT IS THE ANGULAR MAGNIFICATION?

Angular magnification $M = \dfrac{\text{angle subtended by the image at the eye}}{\text{angle subtended by the object at unaided eye}}$

$$M = \frac{\beta}{\alpha} = \frac{f_o}{f_e}$$

where $f_o$ is the focal length of the objective and $f_e$ is the focal length of the eyepiece.

For small angles: $\alpha = \tan \alpha = \dfrac{h}{f_o}$ and $\beta = \tan \beta = \dfrac{h}{f_e}$

- The brightest image is seen when the eye is positioned where an image of the objective is formed by the eyepiece. This is where the **eye-ring** is placed.

### PROBLEMS WITH REFRACTING TELESCOPES

- Astronomical telescopes need large apertures to capture as much light as possible from dim objects.
- It is difficult and expensive to correct lenses of large aperture for **chromatic** and **spherical aberrations**.
- Long telescopes are needed to produce high magnifications.

### PRODUCING A REAL IMAGE

- An image can be focused on a screen or photographic plate by moving the eyepiece further away from the objective lens.

Diameter of aperture of largest refracting telescope $\approx 1\,\text{m}$

# REFLECTING TELESCOPES

## CONCAVE MIRROR

All wavelengths obey the law of reflection $i = r$ so there is **no chromatic aberration**. Silvering is on the front surface to avoid images formed by multiple reflections.

Parallel rays from a distant object form an image at the focus of the mirror.

The focal length is **half the radius of curvature** of the mirror.

- The image is blurred due to **spherical aberration** with large aperture mirrors.
- This is avoided by using a **parabolic mirror**.

## NEWTONIAN REFLECTING TELESCOPE

**Objective mirror** produces image of distant object at the principal focus.

**Eyepiece** produces virtual image at infinity.

Some reflecting telescopes have objective mirrors up to 10 m diameter.

**Plane mirror** reflects rays to form real image to enable viewing using eyepiece.

## CASSEGRAIN TELESCOPE

- Telescope length is reduced using this arrangement:

- The secondary convex lens effectively increases the overall focal length, increasing the magnification.
- Image formed by **primary mirror** at the principal focus is a virtual object for the convex **secondary mirror**.

**Eyepiece** produces virtual image of the object at infinity.

The final real image is formed in position of a small hole in the primary mirror.

Spherical mirrors used because they are easier to make than parabolic mirrors. **Glass correction plate** is used to eliminate defects.

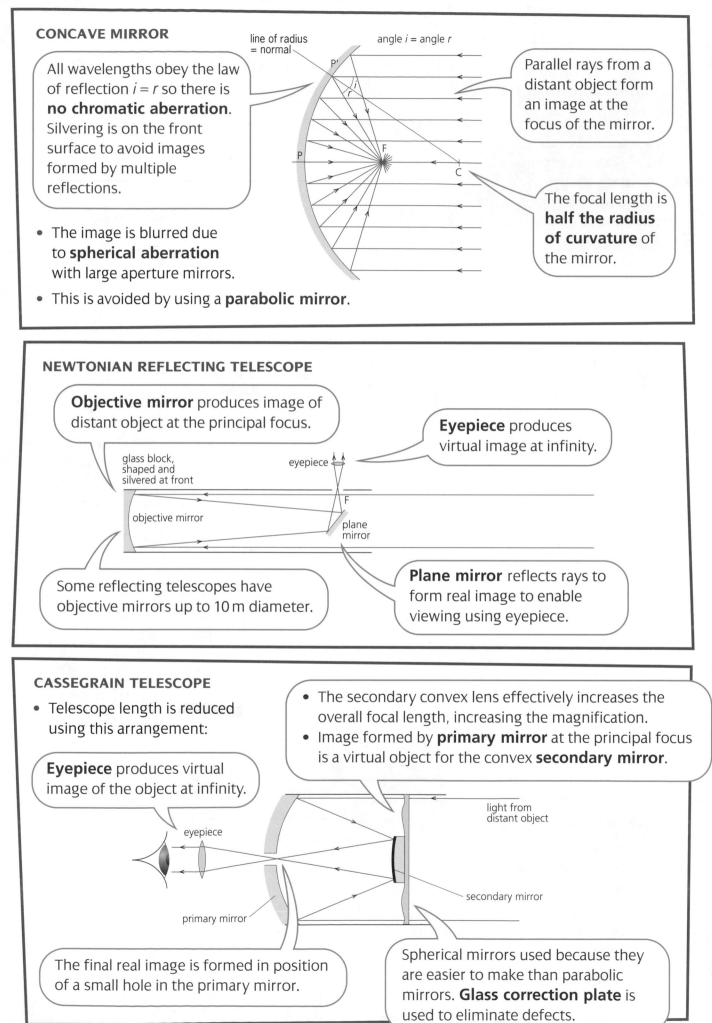

# CHARGE COUPLED DEVICES (CCD)

- A CCD is the digital equivalent of the film inside a conventional camera.

Each **pixel** is a small rectangular piece of light-sensitive semiconductor called a **photosite**.

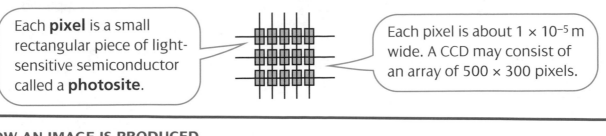

Each pixel is about $1 \times 10^{-5}$ m wide. A CCD may consist of an array of $500 \times 300$ pixels.

## HOW AN IMAGE IS PRODUCED

- Photons of light fall on a photosite and liberate electrons from atoms.
- Brighter light means more photons per second so more electrons are liberated.
- The pixels are insulated so that charge builds up on each pixel.
- The charge built up on each pixel is proportional to the light intensity falling on it.

- After exposure, the charge on each pixel is monitored systematically by an array of wires connected to electronic circuitry.
- The charge levels are stored in a computer.
- The computer converts the charge levels into light levels and produces an image.

## HOW ARE COLOURED IMAGES PRODUCED?

### Method 1
- In astronomy, three images are taken over long intervals with red, green and blue filters. The images can then be processed to produce a final coloured image.

### Method 2
- Digital cameras use a semiconductor wafer in which many sets of three adjacent pixels are covered with red, green and blue filters.
- Information about the brightness and colour in the area covered by the three pixels is stored by the charge that builds up in each pixel.

## WHAT IS MEANT BY QUANTUM EFFICIENCY?

- **Quantum efficiency** is the percentage of incident photons used by the device to form the image.
- For CCDs, the quantum efficiency is 70% compared with 4% for chemical film and 1% for the eye.
- A high quantum efficiency (more photons used):
  - allows shorter exposure times
  - improves the ability to produce brighter images of dim objects.

# LIMITATIONS OF OBSERVATIONS

## DIFFRACTION EFFECTS

- Resolution is limited by diffraction of radiation at circular apertures. (A point source produces circular diffraction pattern called an **Airy disc.**)
- The **Rayleigh criterion** gives the **resolving power** of a circular aperture as:

$$\frac{1.22\lambda}{b}\left(\approx \frac{\lambda}{b}\right)$$

where $\lambda$ = wavelength and $b$ is the diameter of the aperture.

## ATMOSPHERIC EFFECTS

- Movement of air produces variable refraction effects as the light travels through the atmosphere, so image is distorted.
- The atmosphere absorbs some wavelengths (e.g. infrared) so spectrum observed is not the true spectrum.
- Atmosphere adds unwanted radiation (noise) to the signal in parts of the electromagnetic spectrum.
- These effects are reduced by placing telescopes on tops of mountains.

# RADIOTELESCOPES

- A radiotelescope has a similar structure to a reflecting telescope.

aerial at focal point of the dish

**Waves of frequency 30 MHz to 600 GHz from distant objects**

**CONTROL ROOM**
- Signal is amplified.
- Telescope builds up an image by scanning across the source.
- Data is processed by computer.
- Dish is moved to track the source.

Parabolic metal dish focuses waves onto the aerial. More power is needed to detect dim objects.
**Power received ∝ (diameter of dish)²**

Pre-amplifier may be **cooled** to low temperature to minimise noise in amplifier circuitry.

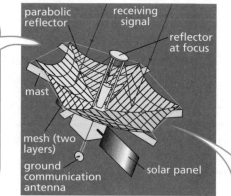

parabolic reflector
receiving signal
reflector at focus
mast
mesh (two layers)
ground communication antenna
solar panel

- Radiotelescopes in **orbit** around the Earth can send data to receiving station.
- This can minimise problems due to:
  - **absorption of radio waves by water vapour** in atmosphere
  - **interference** from electronic equipment used on Earth.
- There can still be problems due to interference from transmissions by other satellites.

- Wire mesh aerial makes dishes lighter so easier to move and cheaper to put into orbit.
- For aerial to be efficient:

$$\text{mesh spacing} \; < \frac{\lambda}{20}$$

where $\lambda$ is the wavelength being studied.

# MAKING OBSERVATIONS USING THE DOPPLER EFFECT

Stars A and B orbit around their centre of mass forming a binary star.

- These are observations of the spectrum when the star is in positions B1, B2 and B3.
- The spectrum is **red-shifted** in position B1 and **blue-shifted** in position B2.

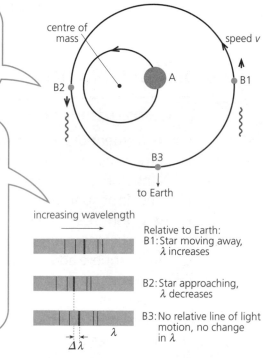

centre of mass
speed v
B2
A
B1
B3
to Earth

increasing wavelength

Relative to Earth:
B1: Star moving away, $\lambda$ increases

B2: Star approaching, $\lambda$ decreases

B3: No relative line of light motion, no change in $\lambda$

$\lambda$
$\Delta\lambda$

- The 'true' wavelength $\lambda$ of a line in the spectrum is observed in position B3.
- $\Delta\lambda$ is the measured shift due to relative motion:

$$\frac{v}{c} = \frac{\Delta\lambda}{\lambda}$$

so $v$ can be found.
- The radius of the orbit can be measured and the period of the orbit found from:

$$T = \frac{2\pi r}{v}$$

- In a similar way, the period of rotation of a star can be found from observations of the spectrum of the light emitted from opposite sides of the star.

# SPECIAL RELATIVITY

## MICHELSON–MORLEY EXPERIMENT

- The experiment demonstrates the **invariance of the speed of light**.
- Michelson and Morley used an **interferometer** to try to detect a change in the speed of light due to the motion of the Earth.
- The experiment attempted to measure the **absolute motion** of the Earth by measuring the difference in the speed of light as the Earth moves through the **aether**.
- As the Earth moved at speed $v$ through the aether, the speed of light was expected to be:
  - $(c - v)$ when the Earth moved in the direction of the light beam
  - $(c + v)$ when the Earth moved in the opposite direction to the light beam.

  (Compare what happens with sound.)

  > $c$ = speed of light through the aether
  > $v$ = speed of the Earth through the aether

- It was thought that a medium called the aether was necessary to carry light waves.

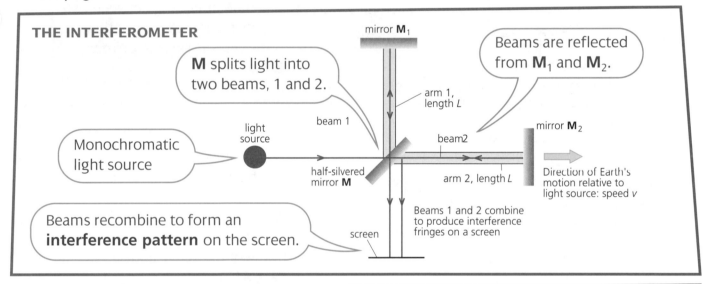

**THE INTERFEROMETER**

mirror $M_1$

**M** splits light into two beams, 1 and 2.

Beams are reflected from $M_1$ and $M_2$.

arm 1, length $L$

beam 1

Monochromatic light source

light source

half-silvered mirror **M**

beam2

arm 2, length $L$

mirror $M_2$

Direction of Earth's motion relative to light source: speed $v$

Beams recombine to form an **interference pattern** on the screen.

screen

Beams 1 and 2 combine to produce interference fringes on a screen

### WHAT RESULT WAS EXPECTED FROM THE EXPERIMENT?

- Beam 1 was perpendicular to motion of Earth so speed is $c$ from **M** to $M_1$ and back.
- Beam 2 was expected to move at $(c - v)$ from **M** to $M_2$ and $(c + v)$ from $M_2$ to **M**.
- The time to travel to the mirror $M_2$ and back should have been longer than to $M_1$ and back.
- The time difference should have introduced an additional phase difference between the two beams.
- When the apparatus was rotated through 90° so that the arms changed places, there should have been an observable shift in the fringe pattern enabling the time difference to be found.

### WHAT WAS THE RESULT AND CONCLUSION OF THE EXPERIMENT?

- The experiment:
  - detected no shift in the interference pattern
  - did not detect any time difference.
- The conclusion was that the speed of light was the same in both directions showing that the speed of light is not affected by the Earth's motion.

> From the known speed of the Earth, they expected to measure times that were **four times greater** than the smallest time measurable by the apparatus.

# EINSTEIN'S THEORY OF SPECIAL RELATIVITY

## POSTULATES OF SPECIAL RELATIVITY

- The invariance (constancy) of the speed of light in free space
- Laws of physics are the same in all **inertial frames of reference**.

- **Inertial frames of reference** move at a **steady speed** relative to each other.
- Objects in inertial frames of reference obey **Newton's first law**.

The time interval between the two events for the moving observer is **longer than** the proper time (time passes more slowly).

## WHAT IS TIME DILATION?

- Einstein showed that the time $t$ between the same two events measured by an observer moving at a speed $v$ relative to the first is given by:

$$t = \sqrt{\frac{t_0}{1 - \frac{v^2}{c^2}}}$$

$$\sqrt{1 - \frac{v^2}{c^2}} < 1 \text{ so } t > t_0$$

- When travelling at $0.83 \times$ the speed of light, time passes twice as slowly.
- The time interval $t_0$ between two events measured by a stationary observer is called the **proper time**.

## ILLUSTRATING TIME DILATION

- Muons have a half-life of $2\,\mu s$.
- Muons created 10 km high in the upper atmosphere travel at $0.998c$ and take $\frac{10\,000}{2.99 \times 10^8} = 33\,\mu s$ ($\approx$ 16 half-lives) to reach the Earth's surface.
- So if there are no relativistic effects, few should be detected at the surface.
- However, in the laboratory frame of reference, half-life is: $\frac{2}{\sqrt{(1 - 0.998^2)}} = 32\,\mu s$
- So only one half-life passes in the journey from upper atmosphere to Earth's surface and about $\frac{1}{3}$ of the muons survive.

## WHAT IS LENGTH CONTRACTION?

- The distance travelled appears much shorter when travelling faster.
- When travelling at a speed $v$ an object of length $l_0$ in a stationary frame of reference appears to have a length:

$$l = l_0\sqrt{1 - \frac{v^2}{c^2}} \text{ or } l_0\left(1 - \frac{v^2}{c^2}\right)^{\frac{1}{2}}$$

For the muon, the Earth's atmosphere appears to have a depth of $10\,000(1 - 0.998^2)^{\frac{1}{2}} = 630\,m$

## HOW DOES MASS CHANGE WITH SPEED?

The increased mass has to be accounted for when:
- measuring charge to mass ratios
- designing particle accelerators.

- If the rest mass of an object is $m_0$ its mass when moving is increased to:

$$m = \frac{m_0}{\sqrt{1 - \frac{v^2}{c^2}}}$$

- The **total energy** of a moving object is given by:

$$E = \frac{m_0 c^2}{\sqrt{1 - \frac{v^2}{c^2}}}$$

Remember that mass and energy are equivalent: $E = mc^2$

The maximum possible speed of a particle is $c$. Mass would be infinite when $v = c$

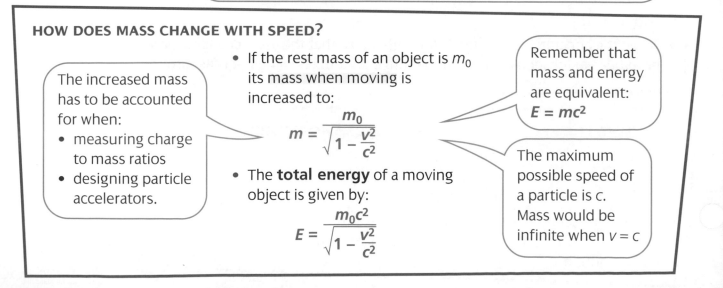

# INDEX